Atlas of Sectional Anatomy

Luciano Alves Favorito • Natasha T. Logsdon
Editors

Atlas of Sectional Anatomy

Understanding the Anatomical Aspects of the Thorax, Abdomen and Pelvis

Springer

Editors
Luciano Alves Favorito
Urogenital Research Unit
Rio de Janeiro State University and
IDOMED Institute
Rio de Janeiro, Rio de Janeiro, Brazil

Natasha T. Logsdon
Urogenital Research Unit, Rio de Janeiro
State University
Geraldo di Biase University Center
Rio de janeiro, Rio de Janeiro, Brazil

ISBN 978-3-030-91687-9 ISBN 978-3-030-91688-6 (eBook)
https://doi.org/10.1007/978-3-030-91688-6

This Springer imprint is published by the registered company Springer Nature Switzerland AG
The registered company address is: Gewerbestrasse 11, 6330 Cham, Switzerland

Preface

Sectional anatomy is the study of normal body structures in sectional planes. Sectional anatomy is the basis for the knowledge of clinical practice, surgery, and imaging exams interpretation as US, TC, and MRI. This book, *Sectional Anatomy of the Thorax, Abdomen and Pelvis*, is an accomplishment of the Urogenital Research Unit.

The Urogenital Research Unit is located in the Biomedical Center at the State University of Rio de Janeiro, Brazil, and under the direction of Francisco J.B. Sampaio is dedicated to advancing the knowledge on the urogenital system, from both a basic science and a medical perspective. Research is conducted on several models, which include clinical trials and experiments on laboratory animals, and employ a wide range of methodologies. The study of the human macroscopic anatomy of the trunk is one of the most important lines of research in this unit.

In this book we present important information and images about the sectional and macroscopic anatomy of trunk applied to clinical practice and image, mainly for radiological specialty of internal medicine. This book is applied to anatomists, general surgeons, clinicians, urologists, gynecologists, pediatric surgeons, PhD researchers, fellows, and medical students, and we hope that it will be useful for clinical practice and medical research.

Rio de Janeiro, RJ, Brazil — Luciano Alves Favorito
Rio de Janeiro, RJ, Brazil — Natasha T. Logsdon

Acknowledgments

The authors would like to thank the National Council for Scientific and Technological Development (CNPq–Brazil), Coordination of Improvement of Higher Education Personnel (CAPES-Brazil), and the Rio de Janeiro State Research Foundation (FAPERJ) for the Urogenital Research Unit Support.

Contents

Contributors

Luciano Alves Favorito, MD, PhD Urogenital Research Unit, Rio de Janeiro State University, Rio de Janeiro, RJ, Brazil

IDOMED-UNESA, Rio de Janeiro, RJ, Brazil

The National Council for Scientific and Technological Development (CNPq), Lago Sul, Federal District, Brazil

The Rio de Janeiro State Research Foundation (FAPERJ), Rio de Janeiro, RJ, Brazil

Natasha T. Logsdon, MSc, PhD University Center Geraldo di Biasi, Rio de Janeiro, RJ, Brazil

Urogenital Research Unit, Rio de Janeiro State University, Rio de Janeiro, RJ, Brazil

Francisco J. B. Sampaio, MD, PhD Urogenital Research Unit, Rio de Janeiro State University, Rio de Janeiro, RJ, Brazil

The National Council for Scientific and Technological Development (CNPq), Lago Sul, Federal District, Brazil

The Rio de Janeiro State Research Foundation (FAPERJ), Rio de Janeiro, RJ, Brazil

Abbreviations

B	Bladder
CMR	Cardiovascular magnetic resonance
CT	Computed tomography
D	Duodenum
H	Heart
IO	Internal obturator muscle
L	Liver
LAM	Levator anus muscle
LK	Left Kidney
LL	Left lung
LV	Lumbar vertebra
LT	Left Testis
MRI	Magnetic resonance images
P	Pancreas
PI-RADS	Prostate Imaging Reporting and Data System
PM	Psoas muscle
RL	Right lung
RK	Right kidney
RT	Right testis
S	Spleen
SI	Small intestine
SVC	Superior vein cava
T	Trachea
TV	Thoracic vertebra
VS	Seminal vesicle

Chapter 1
Sectional Anatomy of the Thorax

Luciano Alves Favorito and Natasha T. Logsdon

1.1 Introduction

The thoracic cavity is divided into the mediastinum and two pleuropulmonary regions. In this chapter, we will make a brief anatomical description of the organs of the thoracic cavity, and we will show a sequence of transverse cuts of the thoracic cavity demonstrating the relationships between the viscera.

1.2 Mediastinum

The mediastinum is located in the thoracic cavity, being the space between the two pleuropulmonary spaces. It has a large amount of loose connective tissue that surrounds its elements and supports them. With advancing age, this connective tissue becomes more rigid, and the viscera of the mediastinum tend to show less mobility (Netter 1978; Williams et al. 1995).

The limits of the mediastinum are as follows: (1) uppermost—upper opening of the chest, formed by the first two ribs, the manubrium of the sternum and the first thoracic vertebrae; (2) inferior—diaphragm muscle; (3) posterior—thoracic spine;

L. A. Favorito (✉)
Urogenital Research Unit, Rio de Janeiro State University, Rio de Janeiro, RJ, Brazil

IDOMED-UNESA, Rio de Janeiro, RJ, Brazil

The National Council for Scientific and Technological Development (CNPq), Lago Sul, Federal District, Brazil

The Rio de Janeiro State Research Foundation (FAPERJ), Rio de Janeiro, RJ, Brazil

N. T. Logsdon
University Center Geraldo di Biasi, Rio de Janeiro, RJ, Brazil

Urogenital Research Unit, Rio de Janeiro State University, Rio de Janeiro, RJ, Brazil

L. A. Favorito, N. T. Logsdon (eds.), *Atlas of Sectional Anatomy*,
https://doi.org/10.1007/978-3-030-91688-6_1

(4) anterior—posterior surface of the sternum; and (5) lateral—parietal pleura (Testut and Jacob 1926; Williams et al. 1995).

Several diseases (especially tumors and cysts) affect the mediastinum in characteristic locations. Its classic division, in regions, is proposed to facilitate the study of these diseases that affect the mediastinal organs. We can divide the mediastinum into four major regions: upper, anterior, middle, and posterior (Netter 1978).

The upper mediastinum is separated from the other portions by an imaginary line that extends from the sternal angle to the level of the intervertebral disc between the fourth and fifth thoracic vertebrae (Fig. 1.1). This imaginary line passes at the bifurcation of the trachea (called carina), which can also be used as an anatomical point for the division of the mediastinum (Testut and Jacob 1926; Bergman et al. 1988).

The lower part of the mediastinum is divided into three additional portions, using the following parameters: (1) anterior mediastinum—located between the posterior surface of the sternum, anteriorly, and the pericardium, posteriorly; (2) medium mediastinum—located between the two layers of the pericardium; and (3) posterior mediastinum—located between the pericardium, anteriorly, and the spine, posteriorly (Fig. 1.1).

The upper mediastinum contains the esophagus and trachea (posteriorly), the thymus or its remnant (anteriorly), and, in an intermediate position, the great vessels related to the thoracic sympathetic nervous chain of the heart, in addition to the vagus and phrenic nerves (Williams et al. 1995). Several diseases can affect this region of the mediastinum. The most frequent are thymomas, teratomas, plunging goiters (growths of the thyroid gland that reach the chest), adenomegalies, aneurysms, and neurogenic tumors (Nakazono et al. 2019).

The anterior mediastinum has thymus and fatty tissue as its main components (Fig. 1.2). In adults, the thymus and its remnants occupy, preferably, the upper portion of the mediastinum. The main tumors that affect this region are thymomas, lipomas, and teratomas. More rarely, the plunging goiter can reach this location.

The main components of the middle mediastinum are the heart and the immediately adjacent portions of the vessels of the base (Fig. 1.3). Knowledge of cardiac anatomy is very important to radiological image interpretation. Current cardiovascular magnetic resonance (CMR) examinations require expert planning, multiple breath holds, and 2D imaging (Moghari et al. 2020). We must emphasize that when analyzing the sectional images, we must take into account the difference in thickness between the right and left ventricles; as we can see in Figs. 1.3 and 1.4, the left ventricle is approximately the triple in thickness of the right ventricle. The main diseases that can affect this region are pericardial cysts, cardiac tumors, bronchogenic cysts, adenomegalies, teratomas, and lymphangiomas (Nakazono et al. 2019).

In the posterior mediastinum, the esophagus, the descending aorta, the azygos veins, the thoracic duct, lymph nodes, and the thoracic sympathetic chain are located. The commonest termination height of the azygos vein in the superior vein cava (SVC) is at the level of the fifth thoracic vertebrae. The anatomy of the azygos system is of very importance as a predictor for higher values of SVC diameter and mediastinum pathology. Such findings can be useful in mediastinal surgery and mediastinoscopy (Koutsoufianiotis et al. 2021). The main conditions that affect this region are aneurysms, esophageal lesions, bronchogenic cysts, neurogenic tumors, adenomegaly, pheochromocytoma, hiatus hernia, and spinal injuries (Fig. 1.5).

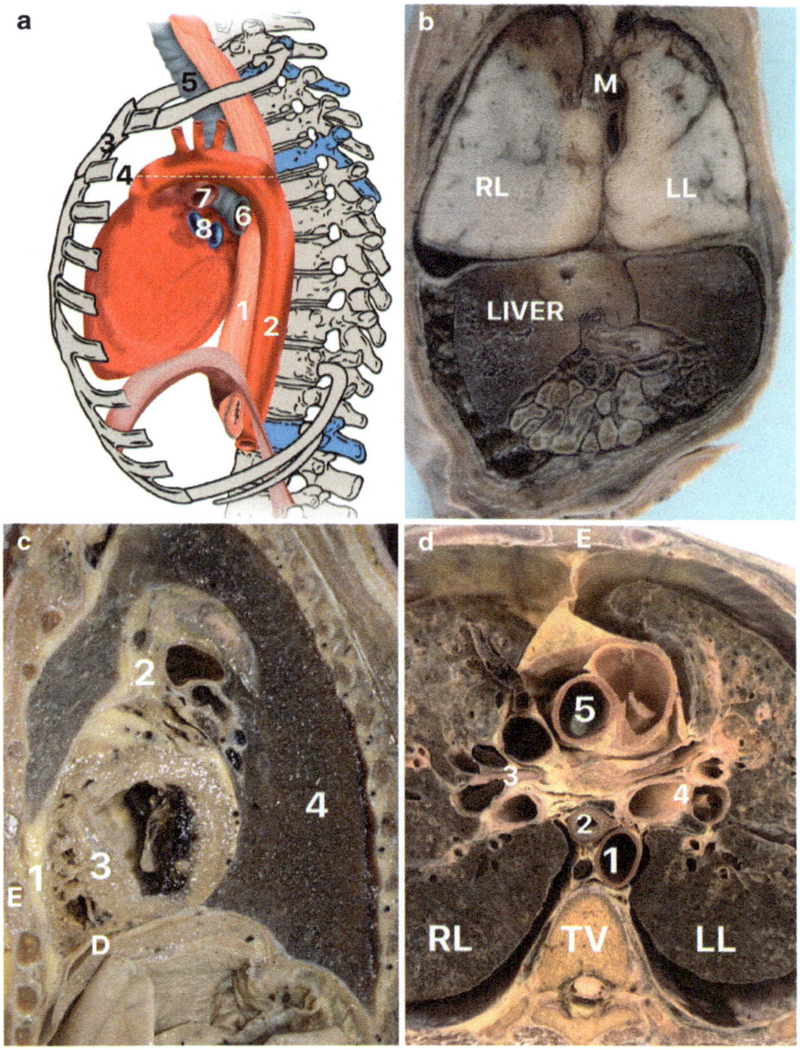

Fig. 1.1 Mediastinum. (**a**) Schematic drawing showing the thoracic cavity and the division of the mediastinum into two compartments by the imaginary line (--------) between the manubrium-sternal junction and the intervertebral body of the fourth thoracic vertebra in superior and inferior mediastinum; 1, esophagus; 2, thoracic aorta artery; 3, sternum; 4, imaginary line that divides the mediastinum; 5, trachea; 6, left main bronchus; 7, left pulmonary artery; 8, left pulmonary vein. (**b**) The figure shows a frontal section in a human frozen fetus in the third gestational trimester showing the relationship between the mediastinum (M) and the right (RL) and left (LL) lungs. (**c**) The figure shows a sagital section of a frozen fresh corpus; we can observe the relationship of the heart (3) with the anterior mediastinal space (1), the lung (4), diaphragm (D), and with the basal vessels (2); E, sternum. (**d**) The figure shows an inferior view of a transverse section of a frozen fresh corpus, at the level of the fourth thoracic vertebra (TV); we can observe some structures of the mediastinum and the relationships with the lungs (RL, right lung; LL, left lung); 1, thoracic aorta artery; 2, esophagus; 3, right pulmonary hilum; 4, left pulmonary hilum; 5, ascending aorta artery

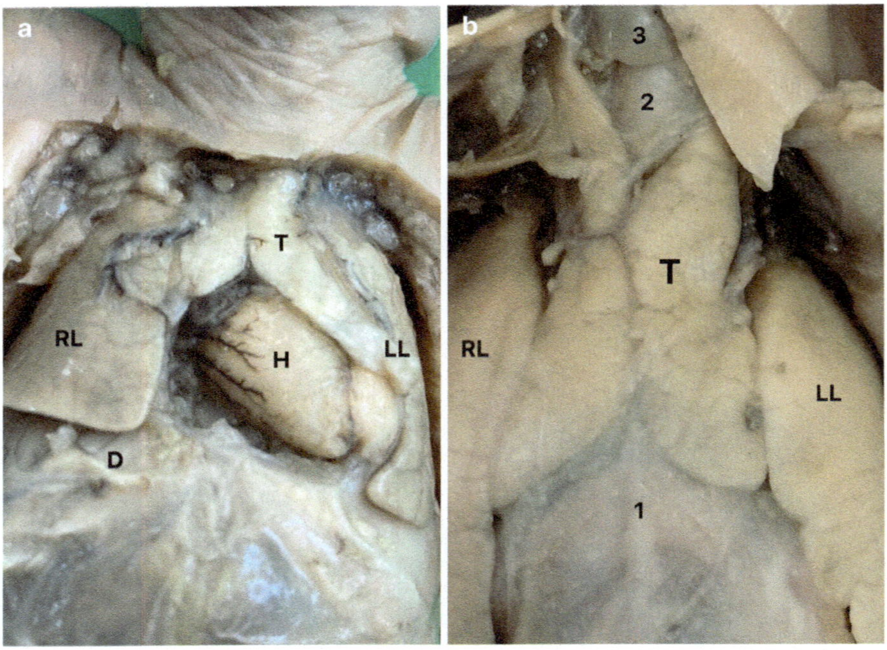

Fig. 1.2 Anterior mediastinum. (**a**) The figure shows the anterior mediastinum of a fixed human fetus in the second gestational trimester; we can observe the timus (T) and the relationships between the heart (H) with the diaphragm (D) and the lungs (RL, right lung; LL, left lung). (**b**) The figure shows the anterior mediastinum of a fixed human fetus in the second gestational trimester; we can observe the timus (T), the pericardium (1), the ascending aorta artery (2), and the tireoid glans (3); RL, right lung; LL, left lung

1.3 Mediastinal Lymph Nodes

Mediastinal lymph nodes can be divided into four main groups: anterior mediastinal, posterior mediastinal, tracheobronchial, and paratracheal (Williams et al. 1995) (Fig. 1.6). Knowledge of their disposition is important due to the diseases that affect them, especially lung cancer.

The posterior mediastinal lymph nodes are located along the esophagus and are responsible for the lymphatic drainage of the intercostal spaces and the parietal pleura. Tracheobronchial lymph nodes are located around the bifurcation of the trachea and along the main bronchi, being divided into the right and left and subcarinal tracheobronchial groups. Subcarinal lymph nodes are of diagnostic importance during the performance of mediastinoscopy and thoracic organ surgeries (Fujiwara et al. 2019). Paratracheal lymph nodes, located along either side of the trachea, drain into the lymph ducts (right and thoracic) and lower cervical lymph nodes (Netter 1978).

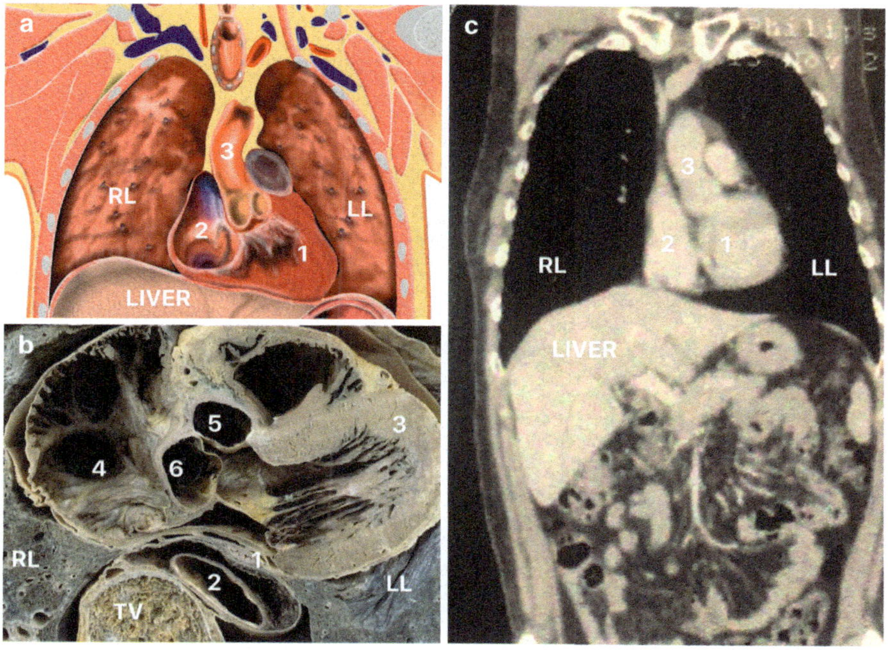

Fig. 1.3 Medium mediastinum. (**a**) Schematic drawing of a frontal section of the thoracic cavity showing some important structures of the medium mediastinum; 1, left ventricle; 2, right atrium; 3, ascending aorta artery; RL, right lung; LL, left lung. (**b**) The figure shows an inferior view of a transverse section of a frozen fresh corpus, at the level of the sixth thoracic vertebra (TV); we can observe some structures of the mediastinum and the relationships with the lungs (RL, right lung; LL, left lung); 1, esophagus; 2, thoracic aorta artery; 3, left ventricle; 4, right ventricle; 5, ascending aorta artery; 6, pulmonary artery. (**c**) The figure shows a thoracic computerized tomography in the frontal section; we can observe the left ventricle (1), the right atrium (2), and the ascending aorta artery (3)

1.4 Pleura and Lungs

1.4.1 Pleura

Pleura is a serous membrane that folds back onto itself to form a two-layered structure that lines the lungs and the inner face of the chest wall. The pleura that surrounds the lungs is called the visceral pleura, and the pleura that is in contact with the chest wall is called the parietal pleura. The two layers are separated by the (virtual) pleural cavity, which is filled with a small amount of pleural fluid. This liquid is essential for sliding between the pleurae and also to prevent the lungs from moving away from the chest wall (Williams et al. 1995).

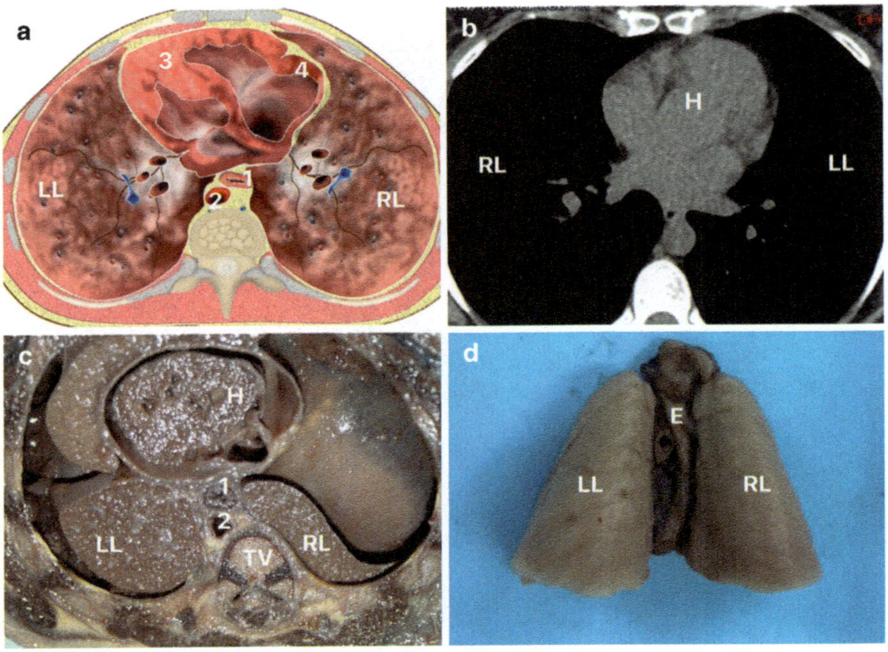

Fig. 1.4 Medium and posterior mediastinum. (**a**) Schematic drawing of a superior view of a transverse section of the thoracic cavity at the level of the sixth thoracic vertebra. We can observe the esophagus (1), the thoracic aorta artery (2), and the relationships between the left ventricle (3) and the right ventricle (4) with the lungs (RL, right lung; LL, left lung). (**b**) The figure shows a thoracic computerized tomography in a transverse section of the thoracic cavity at the level of the sixth thoracic vertebra; we can observe the heart (H) and the lungs (RL, right lung; LL, left lung). (**c**) The figure shows a superior view of a transverse section of the thoracic cavity at the level of the eighth thoracic vertebra. We can observe the esophagus (1), the thoracic aorta artery (2), and the relationships between the heart (H) with the lungs (RL, right lung; LL, left lung). (**d**) In this figure, we can observe the posterior view of thoracic viscera of a human fetus in the second gestational trimester; E, esophagus; LL, left lung; RL, right lung

The visceral pleura follows the divisions of the lungs into lobes, forming pulmonary fissures, and also penetrates the lung parenchyma, dividing the lobes into pulmonary segments. There is no cleavage plane between the visceral pleura and the lung tissue itself. The parietal pleura is separated from the structures of the chest wall by a small amount of connective tissue called endothoracic fascia, and thus the parietal pleura can be easily removed from the chest wall (Netter 1978).

There are four divisions for the parietal pleura according to the area it covers: (1) costal pleura, which lines the ribs and costal cartilages; (2) diaphragmatic pleura, which covers the diaphragm; (3) mediastinal pleura, which is in contact with the mediastinal viscera; and (4) dome of pleura, which lines the pulmonary apex (Mouchova et al. 2018) (Fig. 1.7).

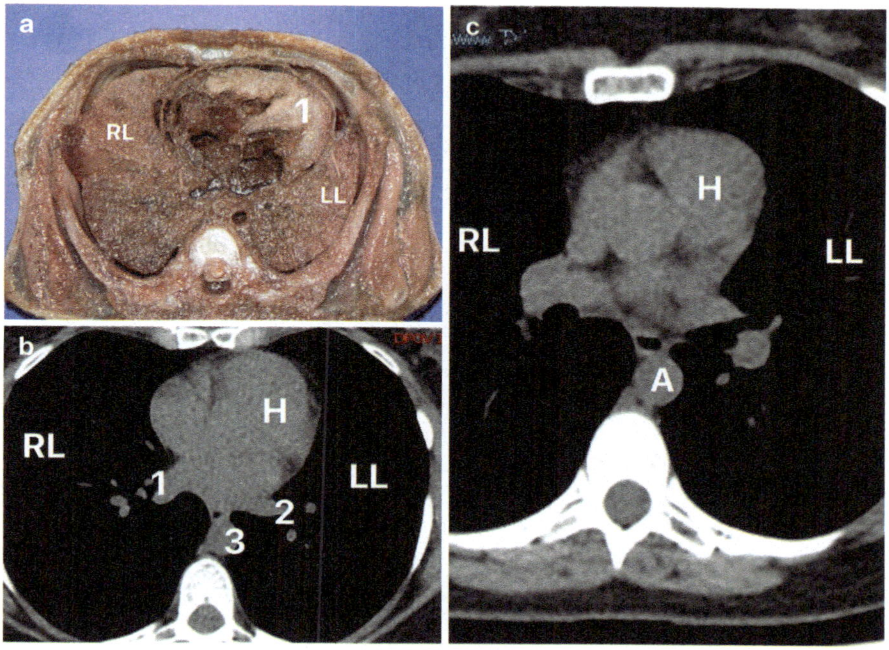

Fig. 1.5 Heart in medium mediastinum. (**a**) The figure shows an inferior view of a transverse section of a frozen fresh fetus in the third gestational trimester at the level of the seventh thoracic vertebra (TV); we can observe the relationships between the heart with the lungs (RL, right lung; LL, left lung). (**b**) The figure shows a thoracic computerized tomography in a transverse section; we can observe the heart (H), left lung (LL), right lung (RL), right pulmonary artery (1), left pulmonary artery (2), and the thoracic aorta artery (3). (**c**) The figure shows a thoracic computerized tomography in a transverse section; we can observe the heart (H), left lung (LL), right lung (RL), and the thoracic aorta artery (A)

The two pleural membranes are continuous in the pulmonary hilum through a pleura cuff that surrounds the structures that enter and leave the lung. Below the root of the lung, the two sides come into contact forming the pulmonary ligament (Williams et al. 1995).

During resting breathing, the expansion of the lungs is not sufficient to fill the entire pleural space. In this way, they form slit-shaped spaces called pleural recesses. There are two pleural recesses: costodiaphragmatic and costomediastinal. The costodiaphragmatic recess is formed between the costal pleura and the diaphragmatic pleura, which are separated only by a capillary layer of pleural fluid. The lower edges of the lungs occupy this space during inhalation; however, on exhalation, the lower edges rise and again allow contact between the two pleural divisions (Fig. 1.8). The same phenomenon is observed in the anterior edges of the lungs that slide in and out of the costomediastinal recesses, during inhalation and exhalation.

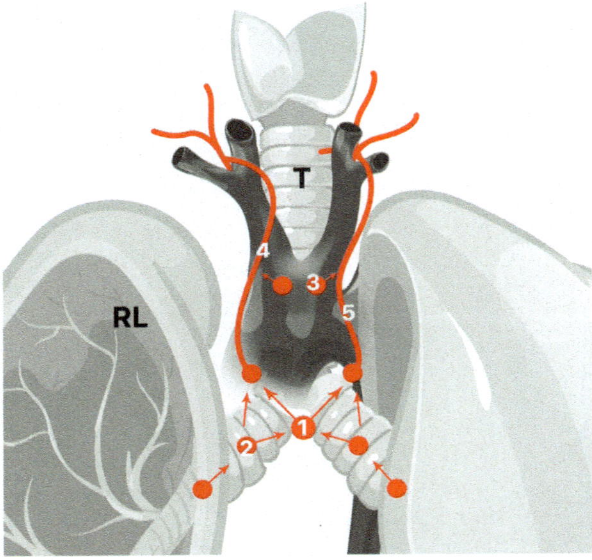

Fig. 1.6 Mediastinal lymph nodes. The figure shows a schematic drawing showing the mediastinal lymph nodes distribution. 1, subcarinal lymph nodes; 2, tracheobronchial lymph nodes; 3, paratracheal lymph nodes; 4, right lymphatic duct; 5, thoracic duct

1.4.2 Lungs

The lungs are conical organs and occupy the lateral regions of the chest cavity from the upper opening of the chest to the diaphragm. However, they are very elastic and reduce their volume to a third or less after opening the chest. They are separated by mediastinal organs such as the trachea, esophagus, heart and major blood vessels (Fig. 1.9).

Each lung has a rounded apex that protrudes into the neck by about 3 cm above the middle part of the clavicle, has a concave base that rests on the diaphragm, and has three faces (costal, diaphragmatic, and mediastinal), which are separated by the anterior margin (separates the mediastinal face from the costal surface anteriorly), posterior margin (separates the mediastinal faces from the costal posteriorly), and inferior margin (separates the diaphragmatic and costal surfaces) (Netter 1978) (Fig. 1.9).

The costal face is large, convex, and related to the costal pleura. The costal face is separated from the ribs, costal cartilages, and intercostal spaces by the costal pleura. The diaphragmatic face is concave and forms the pulmonary base. This concavity is more pronounced in the right lung compared to the left one due to the higher position of the diaphragm on the right because of the liver. The mediastinal face is related to the mediastinal organs, and it is also concave. This concavity is more pronounced on the left, as the heart is shifted two-thirds to the left.

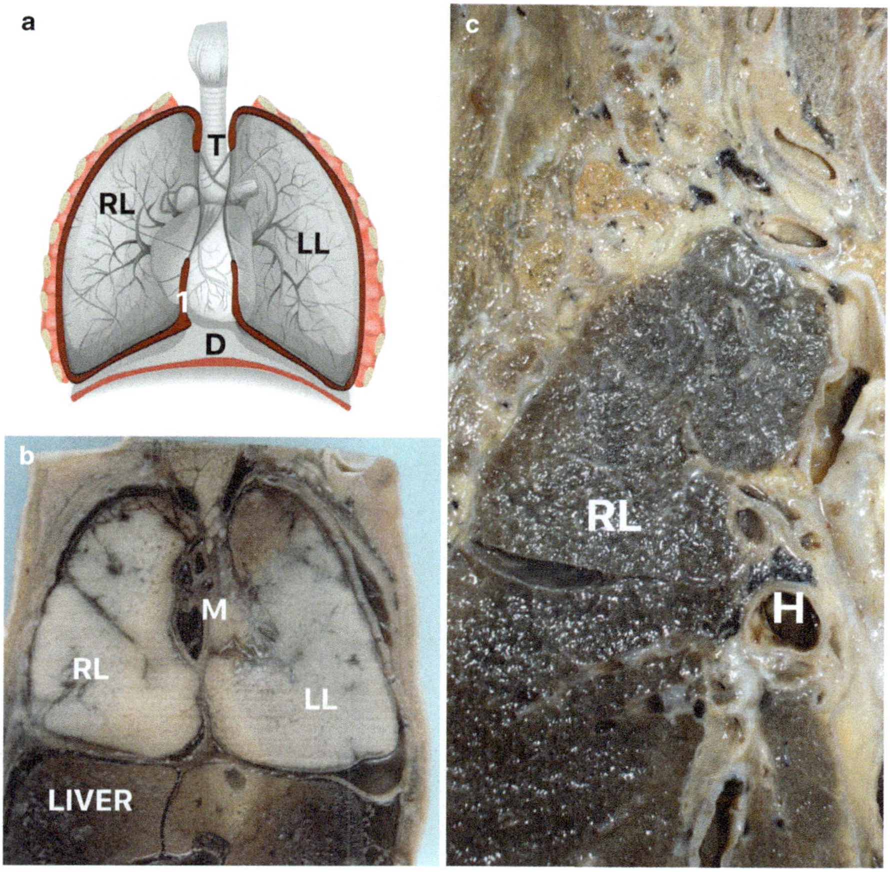

Fig. 1.7 Pleura and pulmonary apex. (**a**) Schematic drawing showing the pleura (1 and all red line) distribution in the thoracic cavity; D, diaphragm; T, trachea; RL, right lung; LL, left lung. (**b**) The figure shows a frontal section of a frozen human fetus in the second gestational trimester; we can oberve the relationships of the mediastinum (M) with the lungs (RL, right lung; LL, left lung). (**c**) Frontal section of a human frozen corpus showing the relationships of the right pulmonary apex; RL, right lung; H, pulmonary hilum

Approximately in the middle of this face is the pulmonary hilum, which is the region where bronchi, vessels, and nerves enter and leave the lung, forming the pulmonary root (Netter 1978; Williams et al. 1995).

The right lung is shorter than the left due to the elevation of the diaphragm. It is also wider as a consequence of the displacement of the heart to the left. Thus, the right lung is heavier, and its total capacity is greater. It has two fissures: horizontal and oblique, and they separate into three lobes (superior, middle, and inferior). The horizontal fissure separates the superior lobe from the middle lobe, and the oblique fissure separates the inferior lobe from the middle and superior lobes. The middle

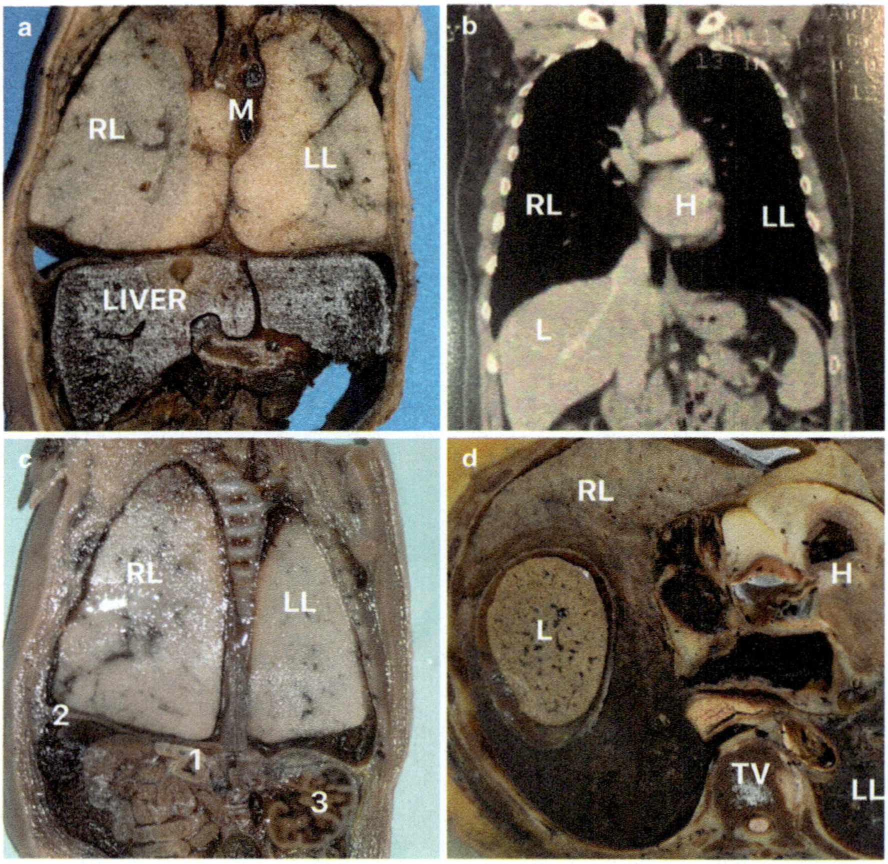

Fig. 1.8 Mediastinum and lungs. (**a**) The figure shows a frontal section of a frozen human fetus in the second gestational trimester; we can oberve the relationships of the mediastinum (M) with the lungs (RL, right lung; LL, left lung). (**b**) The figure shows a thoracic computerized tomography in frontal section; we can observe the heart (H), the right lung (RL), the left lung (LL), and the liver. (**c**) The figure shows a posterior frontal section of a frozen human fetus in the second gestational trimester; we can oberve the lungs (RL, right lung; LL, left lung), the right suprarenal gland (1), the diaphragm (2), and the left kidney (3). (**d**) The figure shows an inferior view of a transverse section of a frozen fresh corpus at the level of the ninth thoracic vertebra (TV); we can observe the heart (H), the lungs (RL, right lung; LL, left lung), and the liver (L)

lobe is wedge-shaped, as it is limited superiorly by the horizontal fissure that is at the level of the fourth costal cartilage and bumps into the oblique fissure in the midaxillary line (Williams et al. 1995).

The left lung has only the oblique fissure, which separates into two lobes: superior and inferior. The lower part of the superior lobe forms an extension between the oblique fissure and the cardiac notch, which is called the lingula, that corresponds to the middle lobe of the right lung.

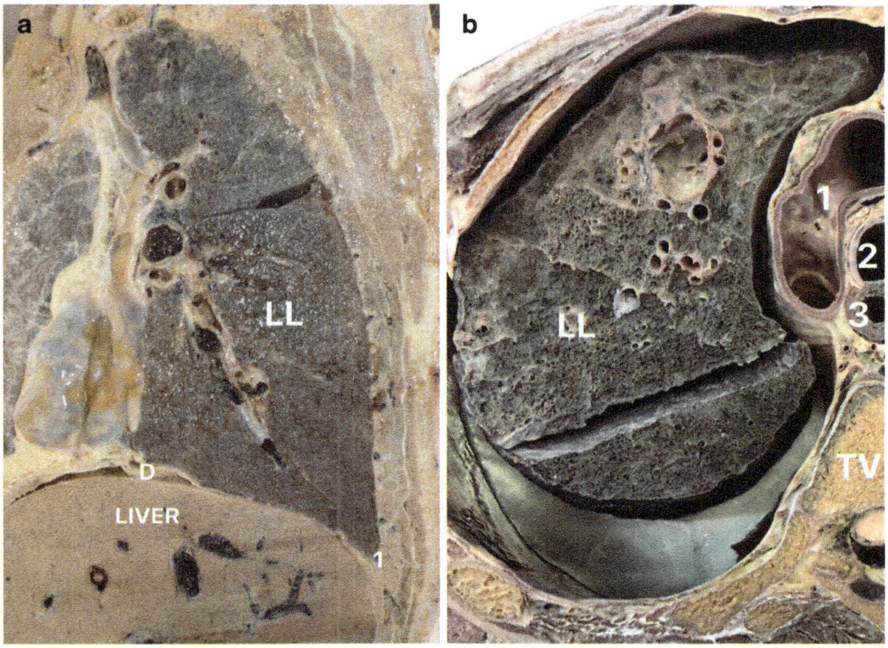

Fig. 1.9 Lungs. (**a**) Frontal section of a human frozen corpus showing the relationships of the left lung (LL) with the diaphragm (D); we can also observe the costophrenic recess (1). (**b**) The figure shows a transverse section of the thoracic cavity of a frozen corpus at the level of the fourth thoracic vertebra (TV); we can observe the left lung (LL), the aortic arch (2), the trachea (2), and the esophagus (3)

1.4.2.1 Bronchopulmonary Segments

Each lobe is divided internally into smaller parts called bronchopulmonary segments. The lung is formed by the union of all bronchial branches; thus, the bronchopulmonary segment is the portion of the lung where a particular bronchus is distributed. The pulmonary segments are the anatomical, functional, and surgical units of the lungs. They are formed by the bronchial tree that begins in the trachea that divides into two main (pulmonary) bronchi. These are divided into three secondary (lobar) bronchi on the right and two on the left, each one for a lobe of the lungs. Finally, the secondary bronchi are divided into tertiary (segmental) bronchi for the pulmonary segments. Knowledge of the lung segment system is essential for understanding human anatomy and has great clinical relevance (Netter 1978). The distribution of the end-branch generation among the five lobes is significantly different. The median branching generation value in the right middle lobe is significantly low compared with that of the other four lobes, whereas that of the right inferior lobe is significantly larger than that of both the right and left superior lobes (Cai et al. 2020; Fujii et al. 2020).

A bronchopulmonary segment is the largest division of the pulmonary lobe. It has a pyramidal shape, with the apex pointing to the pulmonary root and the base facing the costal surface, and it is separated from adjacent segments by septa of connective tissue. It has its own supply through a tertiary bronchus and a segmental artery, which is a tertiary branch of the pulmonary artery and can be surgically removed without affecting the anatomy of nearby segments. The drainage of the bronchopulmonary segment is made by intersegmental veins that drain through the connective tissue that separates one segment from the other. Thus, venous drainage does not respect segmentation, as a vein drains adjacent segments (Netter 1978).

Each tertiary bronchus divides into approximately 20 terminal bronchioles that branch into respiratory bronchioles. These, again, are divided into alveolar ducts, which end in the pulmonary alveoli. The alveolus is the structural unit where the gas exchange takes place.

The bronchopulmonary segments are named according to their position in the lobes. The superior lobe of the right lung has three segments: one apical, one posterior, and one anterior. The right middle lobe has two segments: the lateral and the medial. The inferior lobe of the right lung has five segments: the upper segment and four basilar segments (the medial, anterior, lateral, and posterior basilar segments). The left lung has two segments in the upper part of the superior lobe, the apico-posterior and the anterior, and in the lingula two more segments called the superior lingular and the inferior lingular segments. In the inferior lobe of the left lung, there are the upper segment and three basilar segments, the anteromedial basal, the lateral basal, and the posterior basal (Williams et al. 1995).

Arterial irrigation of the lungs is done by the pulmonary arteries that take blood (venous) to be oxygenated and by the bronchial arteries that irrigate with nonrespiratory parts such as the larger caliber bronchi and the pulmonary support tissue with oxygenated blood. The pulmonary arteries are branches of the pulmonary trunk that branch into two pulmonary arteries, one pulmonary artery for each lung, that enter the pulmonary hiluses and branch out following the division of the bronchi (Saha and Srimani 2019). The left bronchial arteries are branches of the thoracic aorta, and those on the right are branches of the upper posterior intercostal arteries or a left upper bronchial artery (Williams et al. 1995).

The pulmonary veins, two for each lung, drain (arterial) blood from the lungs to the heart. Venous drainage begins in the pulmonary capillaries, with the joining of smaller veins that drain in the intersegmental septa to larger veins. The venous drainage is of the intersegmental type and does not accompany the arterial or bronchial branching. An intersegmental vein drains adjacent segments.

Pulmonary lymphatic drainage is performed by deep lymphatic vessels and superficial lymphatic drainage by subpleural vessels. Superficial drainage is launched in the deep vessels that accompany the bronchi and pulmonary vessels, towards the pulmonary hilum. The final route is made to the tracheobronchial lymph nodes on the same side. Sometimes the inferior lobe of the left lung drains into the lower tracheobronchial lymph nodes on the opposite side (Fig. 1.7).

In Figs. 1.10, 1.11, 1.12, and 1.13, we show a sequence of sectional sections of the chest showing the main structures that can be visualized.

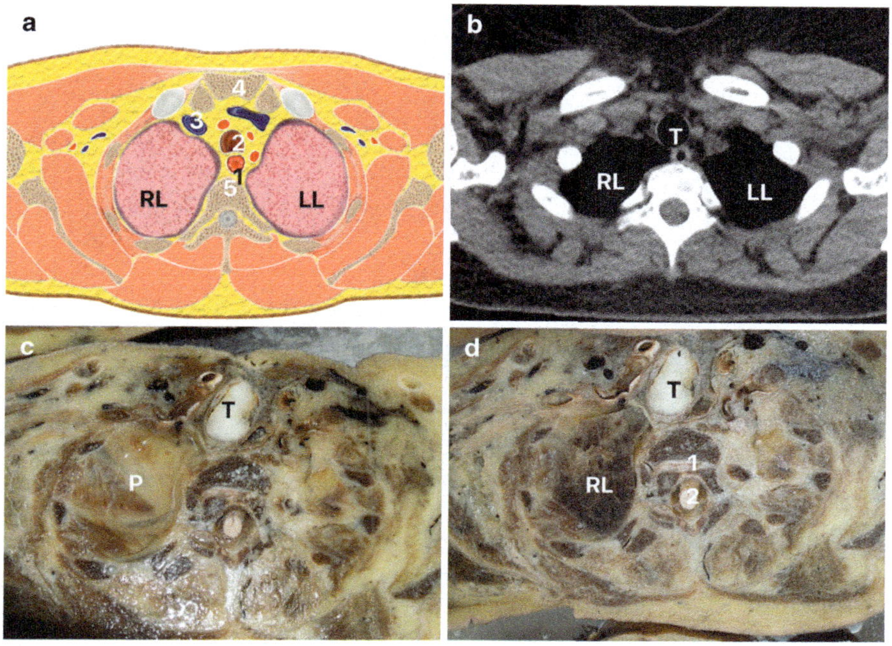

Fig. 1.10 Thoracic sectional anatomy. (**a**) The figure shows a schematic drawing of an inferior view of the thoracic cavity transverse section at the level of the second thoracic vertebra (5); we can observe the right lung (RL), the left lung (LL), esophagus (1), trachea (2), superior vena cava (3), and manubrium of the sternum. (**b**) The figure shows a thoracic computerized tomography in a transverse section at the level of the second thoracic vertebra; we can observe the left lung (LL), right lung (RL), and the trachea (T). (**c**) The figure shows an inferior view of a transverse section of a frozen fresh corpus, at the level of the first thoracic vertebra (TV); we can observe the right pleural cupula (P) and the trachea (T). (**d**) The figure shows an inferior view of a transverse section of a frozen fresh corpus, at the level of the second thoracic vertebra (1); we can observe the right lung (RL) and the trachea (T)

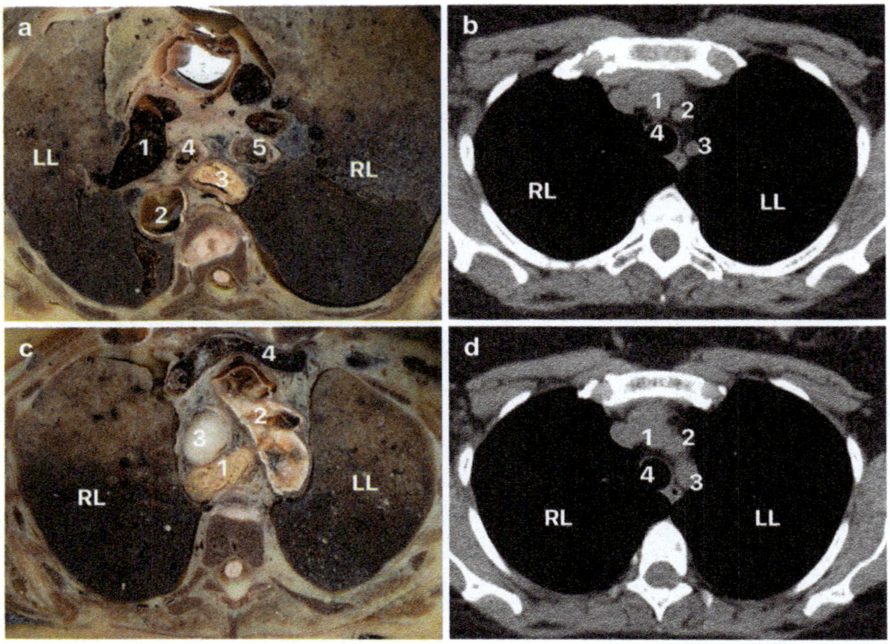

Fig. 1.11 Thoracic sectional anatomy. (**a**) The figure shows a superior view of a transverse section of a frozen fresh corpus, at the level of the fourth thoracic vertebra; we can observe the right lung (RL), the left lung (LL), the left pulmonary artery (1), thoracic aorta artery (2), esophagus (3), left bronchus (4), and right bronchus (5). (**b**) The figure shows a thoracic computerized tomography in a transverse section at the level of the second thoracic vertebra; we can observe the left lung (LL), right lung (RL), the brachiocephalic artery (1), common carotid artery (2), and subclavian artery (3) and trachea (4). (**c**) The figure shows an inferior view of a transverse section of a frozen fresh corpus, at the level of the third thoracic vertebra; we can observe the right lung (RL), the left lung (LL), esophagus (1), aortic arch (2), trachea (3), and the left brachiocephalic vein (4). (**d**) The figure shows a thoracic computerized tomography in a transverse section at the level of the third thoracic vertebra; we can observe the left lung (LL), right lung (RL), the brachiocephalic artery (1), common carotid artery (2), and subclavian artery (3) and trachea (4)

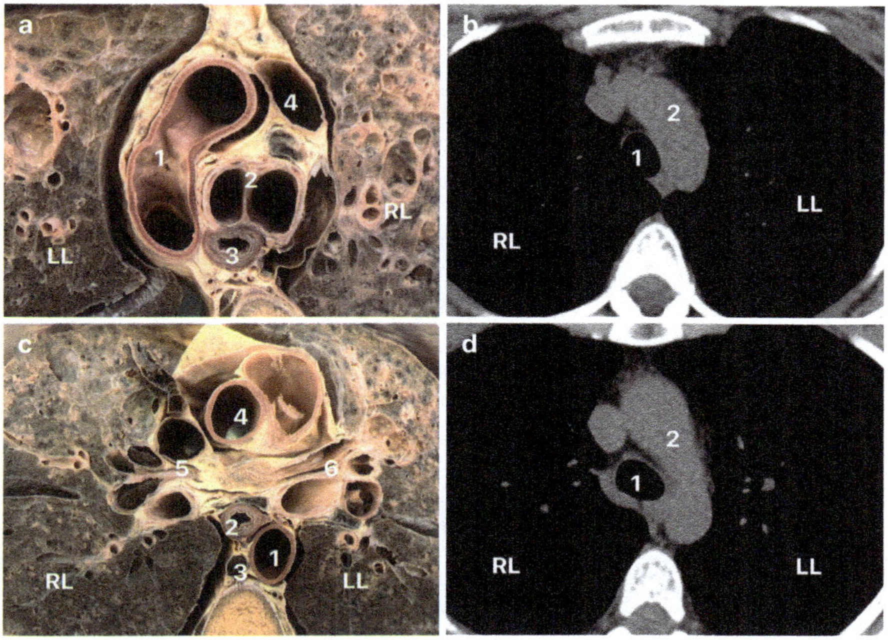

Fig. 1.12 Thoracic sectional anatomy. (**a**) The figure shows a superior view of a transverse section of a frozen fresh corpus, at the level of the fourth thoracic vertebra; we can observe the right lung (RL), the left lung (LL), aortic arch (1), tracheal bifurcation (2), esophagus (3), and pulmonary artery (4). (**b**) The figure shows a thoracic computerized tomography in a transverse section at the level of the transition between the third and fourth thoracic vertebra; we can observe the left lung (LL), right lung (RL), trachea (1), and aortic arch (2). (**c**) The figure shows an inferior view of a transverse section of a frozen fresh corpus, at the level of the transition between the fourth and the fifth thoracic vertebra; we can observe the right lung (RL), the left lung (LL), thoracic aorta artery (1), esophagus (2), azygos vein (3), pulmonary trunk (4), right pulmonary hilum (5), and left pulmonary hilum (6). (**d**) The figure shows a thoracic computerized tomography in a transverse section at the level of the transition between the third and fourth thoracic vertebra; we can observe the left lung (LL), right lung (RL), trachea (1), and aortic arch (2)

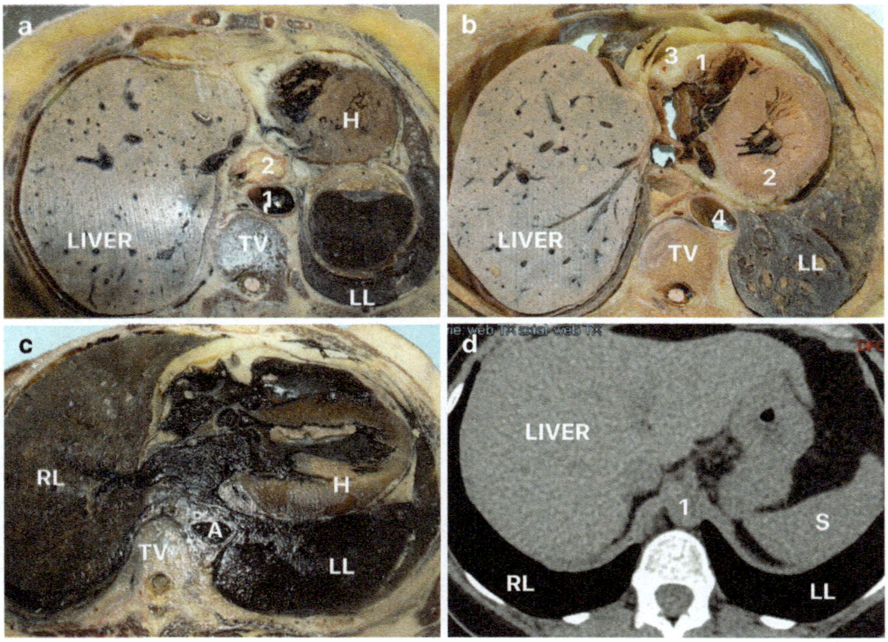

Fig. 1.13 Thoracic sectional anatomy. (**a**) The figure shows an inferior view of a transverse section of a frozen fresh corpus, at the level of the ninth thoracic vertebra (TV); we can observe the liver, the left lung (LL), the heart (H), thoracic aorta artery (1), and esophagus (2). (**b**) The figure shows an inferior view of a transverse section of a frozen fresh corpus, at the level of the eighth thoracic vertebra (TV); we can observe the liver, the left lung (LL), the right ventricle (1), the left ventricle (2), the coronary artery (3), and the thoracic aorta artery (4). (**c**) The figure shows an inferior view of a transverse section of a frozen fresh corpus, at the level of the seventh thoracic vertebra (TV); we can observe the right lung (RL), the left lung (LL), thoracic aorta artery (A), and the heart (H). (**d**) The figure shows a thoracic computerized tomography in a transverse section at the level of the tenth thoracic vertebra; we can observe the left lung (LL), right lung (RL), aorta artery (1), and the spleen (S)

References

Bergman RA, Thompson SA, Afifi AK, Saadeh FA. Compendium of human anatomic variation: catalog, atlas and world literature. Baltimore: Urban & Schwarzenberg; 1988.

Cai P, You Y, Jin ZW, et al. Three-dimensional analysis of the segmental arrangement of lower lung lobes in human fetuses: is this arrangement a miniature version of adult morphology? J Anat. 2020;236(6):1021–34.

Fujii S, Muranaka T, Matsubayashi J, et al. The bronchial tree of the human embryo: an analysis of variations in the bronchial segments. J Anat. 2020;237(2):311–22.

Fujiwara H, Kanamori J, Nakahima Y, et al. An anatomical hypothesis: a "concentric-structured model" for the theoretical understanding of the surgical anatomy in the upper mediastinum required for esophagectomy with radical mediastinal lymph node dissection. Dis Esophagus. 2019;32(8):doy119.

Koutsoufianiotis K, Daniil G, Paraskevas G, et al. Computed tomography angiography study of the azygos vein course and termination into superior vena cava: gender and age impact. Surg Radiol Anat. 2021;43(3):353–61.

Moghari MH, Geest RJ, Brighenti M, Powell AJ. Cardiac magnetic resonance using fused 3D cine and 4D flow sequences: validation of ventricular and blood flow measurements. Magn Reson Imaging. 2020;74:203–12.

Mouchova Z, Rosova B, Matej R. Frozen section of lung, pleura and mediastinum specimen: retrospective analysis of 5-years practical experiences and review of the literature. Cesk Patol. 2018;54(3):127–31.

Nakazono T, Yamaguchi K, Egashira R, et al. CT-based mediastinal compartment classifications and differential diagnosis of mediastinal tumors. Jpn J Radiol. 2019;37(2):117–34.

Netter F. Heart, vol. 2. New Jersey: Ciba; 1978.

Saha A, Srimani P. Comprehensive study of pulmonary hilam with its clinical correlation. Ann Anat. 2019;222:61–9.

Testut L, Jacob O. Anatomia Topográfica. Tomo I. Barcelona: Salvat; 1926. p. 643–781.

Williams PL, Bannister LH, Berry MM, Dyson M, Dussek JE, Ferguson MWJ. In Gray's anatomy. 38th ed. Churchill-Livingstone: Edinburgh; 1995.

Chapter 2
Sectional Anatomy of the Abdomen

Luciano Alves Favorito and Natasha T. Logsdon

2.1 Introduction

The abdominal cavity is divided into two compartments by the transverse mesocolon (peritoneal fold that joins the transverse colon to the posterior wall of the abdomen): supramesocolic compartment and inframesocolic compartment (Hollishead and Rosse 1991; Moore and Dalley 2001) (Fig. 2.1). The supramesocolic compartment contains the liver, the stomach, the first portion of the duodenum, and the spleen. The inframesocolic space contains the small intestine and the large intestine. The greater omentum (large epiploon), which has a large amount of fat, is reflected between the stomach and the transverse colon, occupying a large part of the abdomen's inframesocolic compartment (Fig. 2.1).

Afterward, we will make a brief anatomical description of the organs of the abdominal cavity, and we will show a sequence of transverse sections of the abdominal cavity showing the relationships between the viscera.

L. A. Favorito (✉)
Urogenital Research Unit, Rio de Janeiro State University, Rio de Janeiro, RJ, Brazil

IDOMED-UNESA, Rio de Janeiro, RJ, Brazil

The National Council for Scientific and Technological Development (CNPq), Lago Sul, Federal District, Brazil

The Rio de Janeiro State Research Foundation (FAPERJ), Rio de Janeiro, RJ, Brazil

N. T. Logsdon
University Center Geraldo di Biasi, Rio de Janeiro, RJ, Brazil

Urogenital Research Unit, Rio de Janeiro State University, Rio de Janeiro, RJ, Brazil

L. A. Favorito, N. T. Logsdon (eds.), *Atlas of Sectional Anatomy*,
https://doi.org/10.1007/978-3-030-91688-6_2

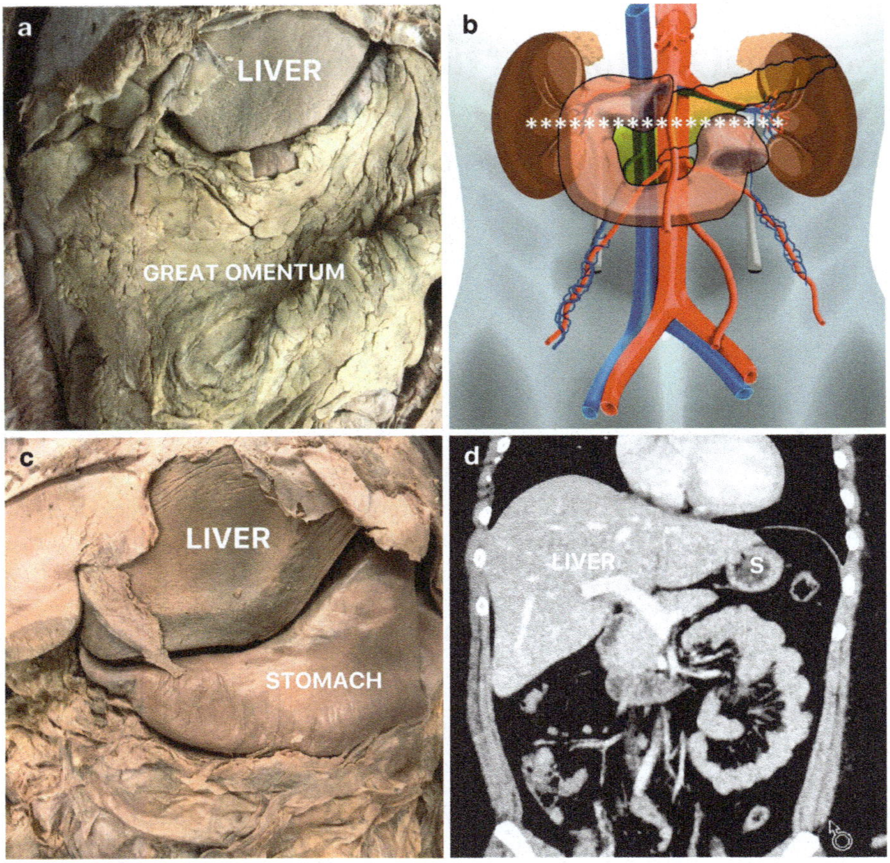

Fig. 2.1 Abdominal cavity. (**a**) The figure shows the abdominal cavity of a fixed corpus; we can observe the great omentum occupying a large part of the inframesocolic compartment of the abdomen. (**b**) Schematic drawing showing the abdominal cavity division into two compartments by the transverse mesocolon position (*********) in the supramesocolic compartment and inframesocolic compartment. (**c**) The figure shows the abdominal cavity of a fixed corpus; we can observe the relationships between the stomach and liver. (**d**) The figure shows a frontal view of an abdominal computerized tomography; we can observe the relationship between the liver and the stomach

2.2 Stomach

The peritoneum completely surrounds the stomach. It leaves the lesser curvature as a lesser omentum (small epiploon) and the greater gastric curvature as a greater omentum (large epiploon). The serous membrane establishes the peritoneal relations of the stomach with the adjacent organs, with the following structures being identified: a) hepatogastric ligament—it is a thickening of the lesser omentum, located at the distal end of the lesser curvature; b) gastrophrenic ligament—located on the posterior face of the stomach to the diaphragm muscle; c) gastrosplenic ligament—located between the greater curvature and the spleen; d) gastrocolic ligament—it is a thickening of the greater omentum projecting from the greater curvature of the stomach to the transverse colon (Williams et al. 1995; Yassi et al. 2010).

The anatomical relationships of the stomach are anterior—the anterior abdominal wall, the left costal margin, the diaphragm, and the left lobe of the liver (Fig. 2.2); and posterior—the omental sac, the diaphragm, the spleen, the left adrenal gland, the upper pole of the left kidney, the splenic artery, the pancreas, and the transverse mesocolon. The topographic relationship between the posterior wall of the stomach and the pancreas favors the endoscopy examination and gastropancreatic transparietal access, which allows the drainage of the cystic collections from the pancreas directly to the gastric lumen (Fig. 2.2) (Netter 1978; Latarjet and Ruiz-Liard 1993).

The omental bursa, also known as the lesser sac, is an irregular space in the subphrenic space, surrounding the caudate lobe of the liver and distributing between the stomach and pancreas, and, although normally collapsed, it can easily be visualized on computed tomography (CT) or magnetic resonance images (MRI) (Fig. 2.3) (Xu et al. 2015).

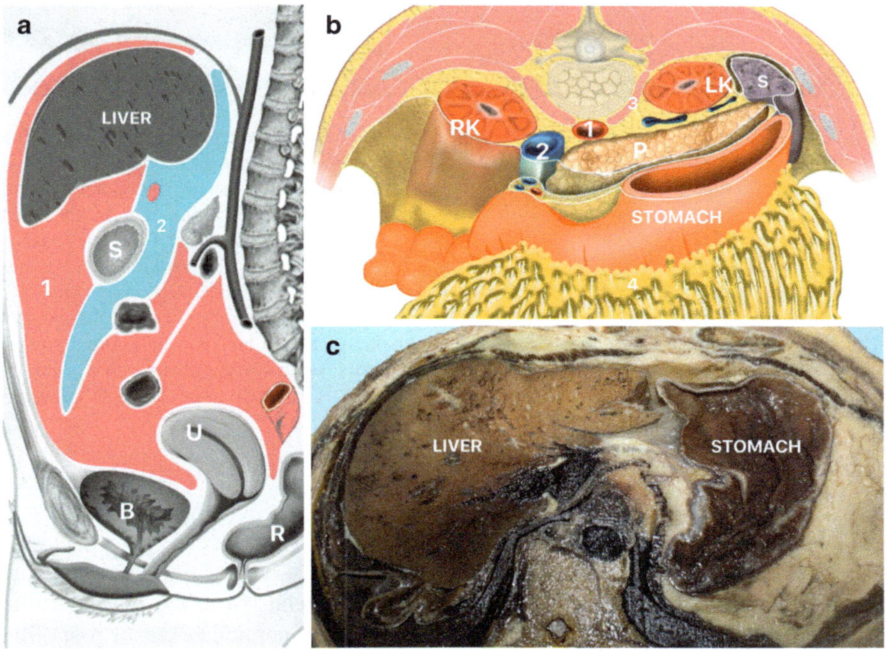

Fig. 2.2 Stomach. (**a**) Schematic drawing of a sagittal section of the abdominal cavity showing the omental bursa (2) position. In this figure, we can observe the relationships between the stomach (S) and liver; 1, abdominal cavity; B, bladder; U, uterus; R, rectum. (**b**) Schematic drawing of a transverse section of the abdominal cavity showing the stomach relationships. We can observe the pancreas (P), spleen (S), right kidney (RK), left kidney (LK), aorta artery (1), inferior vena cava (2), diaphragmatic pillar (3), and great omentum (4). (**c**) The figure shows an inferior view of a transverse section at the level of the ninth thoracic vertebra in a frozen corpus. We can observe the relationships between the stomach with the liver

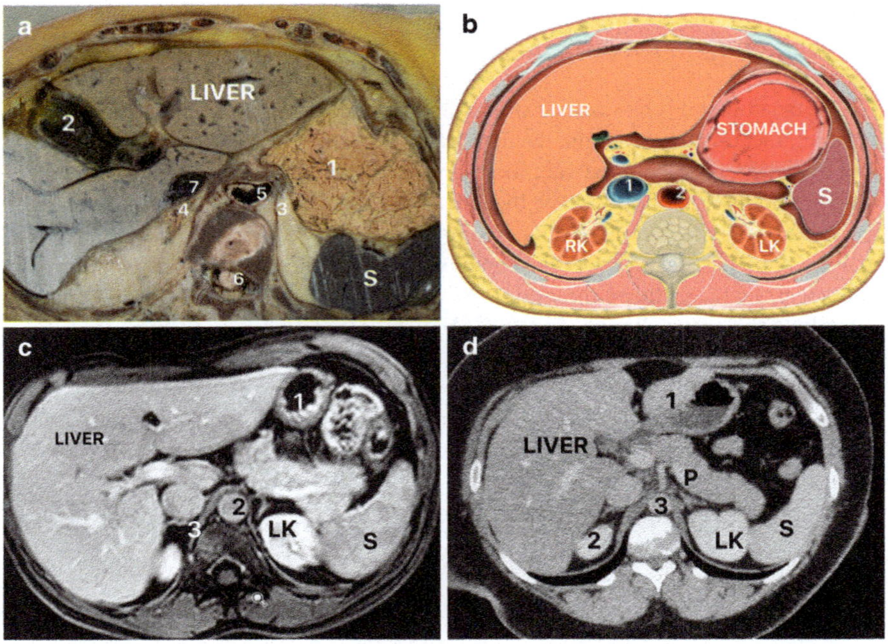

Fig. 2.3 Stomach. (**a**) The figure shows an inferior view of a transverse section at the level of the tenth thoracic vertebra in a frozen corpus. We can observe the relationships between the stomach (1) with the spleen (S) and the liver. 2, gallbladder; 3, diaphragmatic pillar; 4, right adrenal gland; 5, aorta artery; 6, spinal medulla; 7, inferior vena cava. (**b**) Schematic drawing of an inferior view of a transverse section at the level of the first lumbar vertebra; we can observe the relationships between the stomach with spleen (S) and the liver; RK, right kidney; LK, left kidney; 1, inferior vena cava; 2, aorta artery. (**c**) The figure shows an abdominal computerized tomography at the level of the 11th thoracic vertebra; we can observe the relationship between the liver and the stomach (1). LK, left kidney; S, spleen; 2, aorta artery; 3, diaphragmatic pillar. (**d**) The figure shows an abdominal computerized tomography at the level of the first lumbar vertebra; we can observe the relationship of the stomach (1) with the liver and pancreas (P). LK, left kidney; 2, right kidney; 3, aorta artery; S, spleen

2.3 Duodenum

The duodenum is the first, the largest, and the smallest part of the small intestine, located between the stomach and the jejunum; it extends from the pylorus to the duodenojejunal flexure. The duodenum occupies the upper and posterior part of the abdominal cavity at the level and anteriorly to the body of the L1, L2 vertebrae, between the pylorus and the duodenojejunal angle. In the proximal segment, it is more superficial than its meso-distal part, which is deeper. It can vary ordinarily between the horizontal (transverse) lines that pass below the level of the navel and superiorly through the anterior end of the eighth rib bilaterally. The duodenum and pancreas are projected onto the posterior abdominal wall by rotation of the intestine and become fixed there by the fusion of its peritoneal lining and mesentery with the

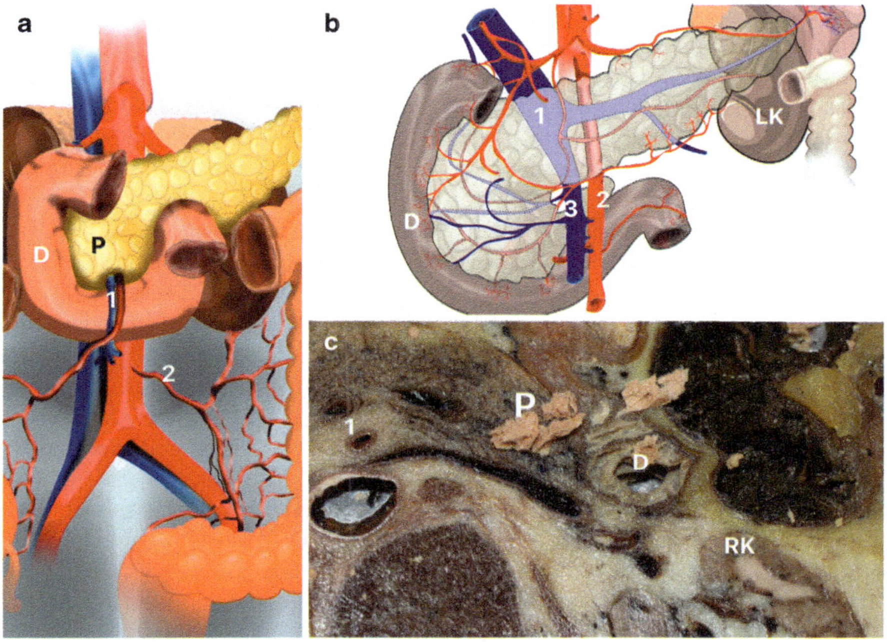

Fig. 2.4 Duodenum. (**a**) Schematic drawing showing the relationship between the pancreas (P) and duodenum (D); we can also observe the superior mesenteric vessels (1) and the inferior mesenteric artery (2). (**b**) Schematic drawing showing the pancreatic and duodenum vascularization; we can observe the hepatic portal vein (1), superior mesenteric artery (2), superior mesenteric vein (3), and the left kidney. (**c**) The figure shows a transverse section of a superior view at the level of the second lumbar vertebra in a frozen corpus; we can observe the duodenum (D) and the pancreas (P); RK, right kidney; 1, superior mesenteric vessels

parietal peritoneum (Fig. 2.4). Thus, only the anterior face of the duodenum and pancreas is covered with peritoneum (later, they are devoid of the peritoneum) (Netter 1978).

First Duodenal Portion (Superior/Bulb) At the beginning, the first portion or upper portion of the duodenum presents a dilation, the duodenal ampulla (duodenal bulb) with a distinct appearance, in which the pylorus protrudes. The remainder leans superiorly and posteriorly along the right side of L1's vertebral body. It is short and has a measurement similar to that of the third part (horizontal), approximately 5 cm in length. The first portion of the duodenum has a considerable degree of mobility that does not exist in the most distal portions. Posterior to the first part of the duodenum, the portal vein, choledochal duct, gastroduodenal artery, and inferior vena cava are located (Bergman et al. 1988). Anatomical relationships: (1) anterior—the liver (square lobe) covers the first anterior portion and the gallbladder is in direct contact with it; (2) posterior—pancreas, portal vein, inferior vena cava, choledochal duct, and gastroduodenal artery, all cross it vertically, being in partial contact; (3) superior—gallbladder neck; and (4) inferior—cervix of the pancreas.

Second Duodenal Portion (Descending) The descending part of the duodenum forms with the preceding one (first part), an angle of approximately 80 degrees: the superior duodenal flexure. It is vertical, is the longest part of the duodenum, and measures 7–10 cm in length. Initially, it is located on the right, parallel to the inferior vena cava, runs inferiorly along the right side of the bodies of the L1, L2, and L3 vertebrae, curving around the head of the pancreas. The main choledochal and pancreatic ducts (from Wirsung) enter its posteromedial wall before perforating the duodenal wall and opening into their lumen at the apex of the major duodenal papilla (papilla of Vater). The minor duodenal papilla, located higher up, can mark the opening in the duodenum of the accessory pancreatic duct. These papillae are usually hidden by circular mucosal folds and present throughout the small intestine. The choledochal duct ends below, perforating the medial wall of the second (descending) part of the duodenum, approximately halfway down the length of the duodenum. It usually joins with the main pancreatic duct and, together, they open in a small dilation, the hepatopancreatic ampulla (by Vater). The ampulla opens in the duodenal lumen through the major duodenal papilla. The terminal parts of both ducts and the ampulla are surrounded by circular muscular extracts known as the muscle or sphincter of the hepatopancreatic ampulla (from Glisson-Oddi). The hilum of the right kidney and its vessels are located on its posterior face. The anterior face is covered by the peritoneum, except along the fixation line of the transverse mesocolon, which crosses the second portion of the duodenum at its midpoint. The descending duodenum is covered above the transverse mesocolon by the liver and below it by the transverse neck and jejunum loops. Anatomical relationships: (1) anterior—transverse neck, transverse mesocolon, and small bowel loops; (2) posterior—hilum of the right kidney, renal vessels, ureter, and psoas major muscle; and (3) medial—head of the pancreas, choledochal duct, and main and accessory pancreatic duct (Netter 1978; Bergman et al. 1988).

Third Duodenal Portion (Horizontal) The horizontal portion starts at the lower duodenal flexure, at a 90-degree angle, and is 6–8 cm long. This part of the duodenum crosses to the left in front of the L3 vertebra and then passes in front of the inferior vena cava and the aortic artery. The horizontal part is later concave to fit the aortomesenteric angle. In front of the aorta artery, it becomes continuous with the short fourth portion, or ascending portion, which returns to the level of L2 vertebra, to the left of the aorta artery, where the duodenojejunal flexure marks its junction with the jejunum (8). Anatomical relationships: (1) anterior—superior mesenteric vessels, loops of the small intestine, and peritoneum; (2) posterior—right psoas major muscle, inferior vena cava, aorta artery, and right ureter; and (3) superior—head and uncinate process of the pancreas and superior mesenteric vessels.

Fourth Duodenal Portion (Ascending) It is as short as the first part of the duodenum, measuring 5 cm in length. It begins at the left of the level of L3 vertebra and ascends to the upper margin of the L2. Soon it ends in the duodenojejunal flexure. This flexure is held in place by a peritoneal fibromuscular fold, the duodenum suspensor muscle (ligament of Treitz). This muscle originates variably from one or all portions of the duodenum, with the exception of the first part, and attaches to the right pillar of the diaphragm. Variations in the length of the duodenal suspensor muscle can over angulate the duodenojejunal flexure and constitute an obstacle, especially if the ligament is short. The union of the fourth part of the duodenum with the jejunum forms an acute angle, usually located at the origin of the mesentery, but variations may occur. It surrounds the head of the pancreas and forms a camber commonly called the duodenal arch. Anatomical relationships: (1) anterior—transverse colon; (2) posterior—left psoas major muscle and left aorta artery margin; (3) medial—head of the pancreas; and (4) superior—beginning of the mesentery root and greater curvature of the stomach (Testut and Jacob 1926; Williams et al. 1995).

2.4 Jejunum and Ileum

The small intestine is one of the most important organs of the digestive system, being responsible for food absorption. It is divided into three parts: duodenum, jejunum, and ileum. The duodenum has been described previously. In this segment, we will cover the anatomy of the jejunum and the ileum, which are the largest portions of the small intestine, 6–7 meters long. The small intestine is located at the inframesocolic space in the abdomen, flanked by the large intestine. Its long portions are accommodated in a relatively narrow region, and these segments are flexed, forming the intestinal loops. The jejunum begins in the duodenojejunal flexure, where the suspensory ligament of the duodenum (ligament of Treitz) is located, while the ileum ends in the region of the ileocecal valve. The duodenojejunal flexure (angle of Treitz) is clinically important, as all bleeding that occurs above it is considered upper digestive bleeding, whereas bleeding that occurs below it is considered lower digestive bleeding.

There is no precise anatomical point that marks the end of the jejunum and the beginning of the ileum. The jejunum corresponds to the proximal 2/3 of the small intestine, while the ileum to the distal 3/5. Most of the jejunum is located in the upper right quadrant of the abdomen, while most of the ileum is located in the lower left quadrant. These two segments of the small intestine are surrounded by a fold of the peritoneum that joins them to the posterior abdominal wall. This portion of the peritoneum is the mesentery. The mesentery is attached to the posterior abdominal wall by the root of the mesentery, which is about 10–12 cm long, beginning at the angle of Treitz and ending at the ileocecal valve. The mesentery root successively crosses the duodenum (ascending and horizontal portions), the abdominal aorta, the inferior vena cava, the right ureter, and the right gonadal vessels (Netter 1978; Williams et al. 1995). The mesentery gives great mobility to the jejunum and ileum and is the place where the intestinal vessels and nerves arrive. The presence of the mesentery causes the intestinal loops to have two faces: the mesenteric side and the anti-mesenteric side. The mesenteric side, where the vessels and nerves arrive, is a region that must be handled with great care to avoid the devascularization of the intestinal loops during surgical procedures. Other important differences between the jejunum and the ileum are the following: the jejunum has a thicker and broader wall, it is more reddish in color, it has a greater amount of folds in its mucosa (also called connective valves or Kerckring's folds), and it presents its meso with less fat, less vascular arches than the mesoileum, its straight and longer vessels, and few aggregated lymph nodes (Peyer plaques) (Fig. 2.5).

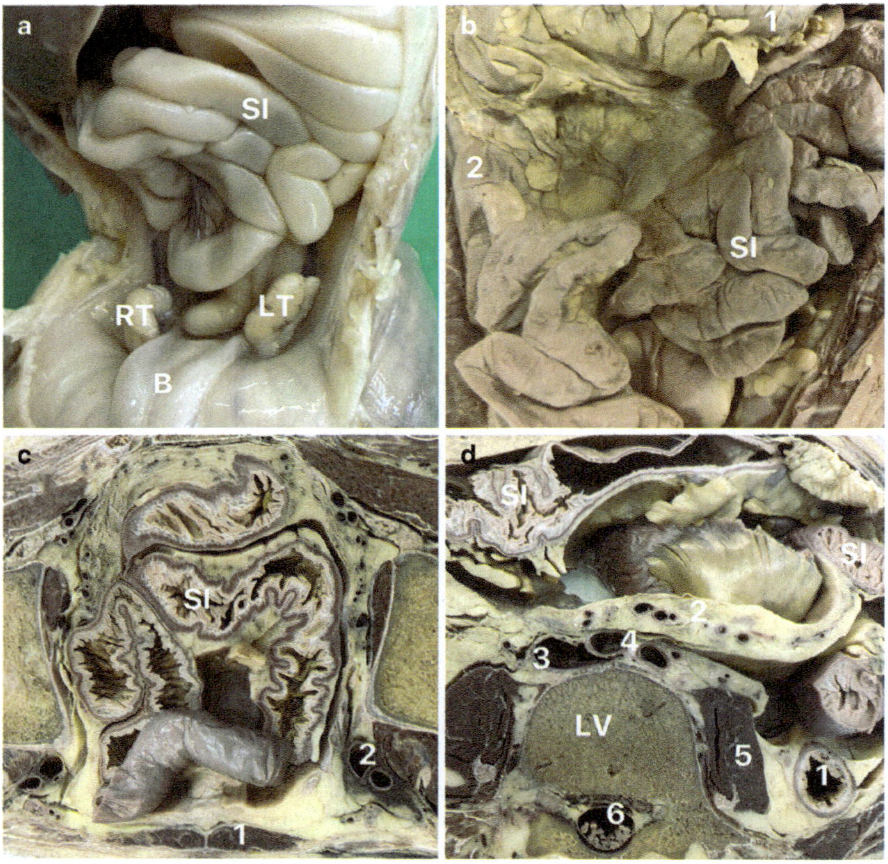

Fig. 2.5 Small intestine. (**a**) The figure shows the abdominal cavity of a human male fetus of the second gestational trimester; we can observe the small intestine (SI). RT, right testis; LT, left testis; B, bladder. (**b**) The figure shows the abdominal cavity of a formalized corpus; we can observe the relationships between the small intestine (SI) and the right colon (1) and the transverse colon (2). (**c**) The figure shows a superior view of a transverse section at the level of the iliac crest in a frozen corpus. We can observe the small intestine (SI). 1, rectus abdominal muscle; 2, external iliac vessels. (**d**) The figure shows an inferior view of a transverse section at the level of the third lumbar vertebra (LV) in a frozen corpus. We can observe the small intestine (SI). 1, left colon; 2, mesenteric root; 3, inferior vena cava; 4, abdominal aorta artery; 5, psoas major muscle; 6, spinal medulla

The arteries that supply the jejunum and ileum are the jejunal and ileal arteries, which arise from the convex edge of the superior mesenteric artery. When these arteries proceed to the loops, they form vascular arches of variable numbers from which the vasa rectae originate. The obstruction of these vessels can occur due to arterial embolisms, especially in patients with cardiac arrhythmias, resulting in a very serious condition, which is the enteric-mesenteric ischemia (Zhou et al. 2006). A prominent place in arterial irrigation of the small intestine is the terminal ileum. The final portion of the ileum is irrigated by the final portion of the superior mesenteric artery and the ileal artery, which is a branch of the ileocolic artery. The anastomosis point between the final part of the superior mesenteric and the ileal arteries is approximately 2 cm from the ileocecal valve. This region has a smaller number of vascular arches, being considered an area of reduced vascularization. This area is called the avascular area of the terminal ileum (avascular space of Treves), and surgeries with resection and anastomosis performed in this region should be avoided due to the high rate of complications due to this poor vascularization.

2.5 Cecum and Appendix

The cecum and appendix belong to the large intestine. However, due to the frequency and importance of diseases in this region, these segments will be described separately. The cecum is a blind bottom pouch, the first portion of the large intestine and the largest one, measuring approximately 7.5 cm in width and 6.5–7.0 cm in length (Williams et al. 1995). It is located in the right iliac fossa, but due to its embryological origin, it may have a more cranial location. It is covered by peritoneum along its entire length, but it does not have mesentery or epiploic appendages. The existence of mesentery in the cecum (the mesocecum) is uncommon, and when it occurs, it gives an excessive mobility to this region, which can lead to its twisting, a very rare condition called cecal volvulus. The shape of the cecum is variable, but we can observe, generally, four types: (1) conical shape, with the appendix originating from its apex—it is rare and corresponds to less than 2% of cases; (2) quadrilateral in shape, with the presence of two saccations and the appendix originating between these sacs—this type is also rare; (3) the third type that is the most common (about 90% of cases) is also composed of two bags, however, of different sizes—the right pouch grows more than the left, cranially, displacing the left pouch inferiorly, arising a new apex, close to the ileocecal valve, where the vermiform appendix originates and where the three taeniae coli converge; and (4) the fourth type, a variation of the third, with an even larger right pouch (Arifuzzaman et al. 2017). The ileocecal valve marks the end of the small intestine, located in the medial portion of the large intestine, at the junction of the cecum with the ascending colon. The ileocecal valve has an ostium and two elevations, called fold flaps; they are two lateral projections of the frenulum of the ileocecal valve. This region, also called the cecal papilla, is surrounded by a musculature that works like a sphincter, controlling the passage of the alimentary bolus from the small intestine to the cecum. The vermiform appendix is a tubular organ on a blind bottom, of a narrow lumen, with a large amount of lymphoid tissue on its wall. It originates at the apex of the cecum, in the region of the confluence of the three taeniae coli. It has a variable size, with an average of 9–10 cm in length. It is surrounded by a meso, the mesoappendix, which gives it relative mobility (Netter 1978) (Fig. 2.6).

Its position is also very variable, located most of the time in the posteromedial region of the cecum, but it can, however, be located behind the cecum or behind the

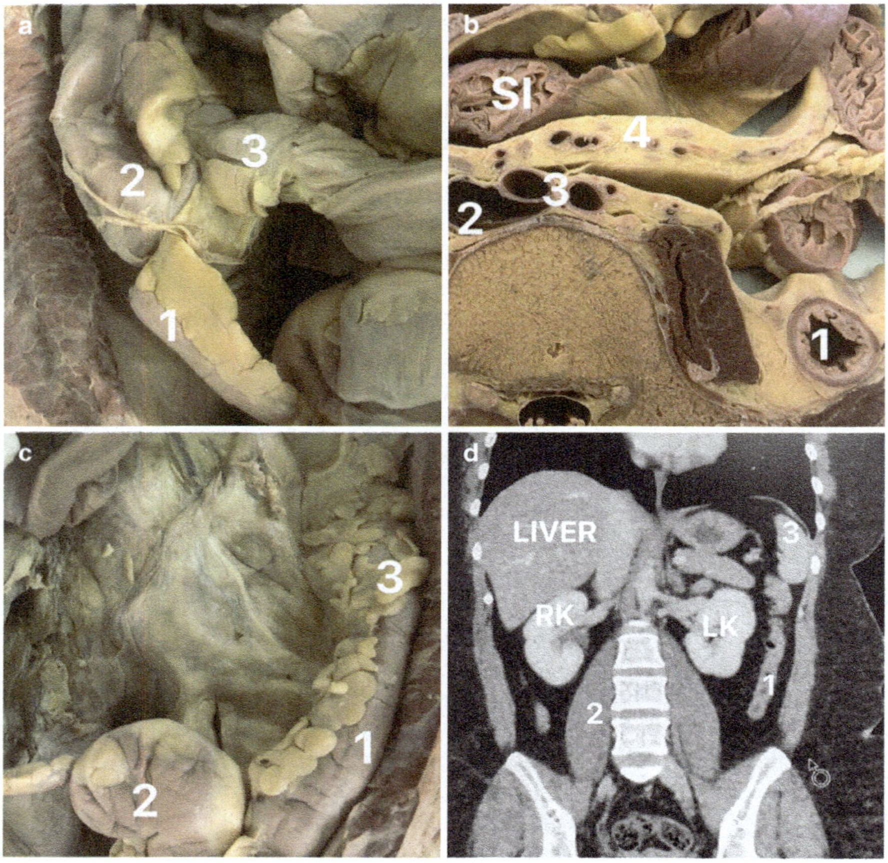

Fig. 2.6 Large intestine. (**a**) The figure shows the abdominal cavity of a formalized corpus; we can observe the relationships between the cecum (1), vermiform appendix (2), and terminal ileum (3). (**b**) The figure shows an inferior view of a transverse section at the level of the third lumbar vertebra in a frozen corpus. We can observe the small intestine (SI), the left colon (1), the Inferior vena cava (2), the bifurcation of abdominal aorta artery (3), and the mesenteric root (4). (**c**) The figure shows the abdominal cavity of a formalized corpus; we can observe the descending colon (1), the sigmoid colon (2), and the epiploic appendix (3). (**d**) The figure shows an abdominal computerized tomography in a frontal section showing the relationships of the left colon (1). RK, right kidney; LK, left kidney; 2, psoas major muscle; 3, spleen

ileum. Knowledge of the different positions of the appendix is of great importance in surgeries for organ resection. The projection of the appendix on the anterior wall of the abdomen has application in a surgical clinic. In most cases, the appendix is projected in the right iliac fossa. If an imaginary line is drawn between the antero-superior iliac spine and the navel, the appendix is located at the junction of the lateral third with the medial two-thirds of this imaginary line, and this region is called the appendicular point or the McBurney point. The cecum and appendix region are irrigated by the ileocolic artery, a branch of the superior mesenteric. The ileocolic artery originates the following branches in this region: anterior cecal artery, posterior cecal, appendicular, ascending cecal, and ileal branches. The appendicular artery is responsible for irrigating the appendix, passing through the mesoappendix and irrigating the organ. The knowledge of its path and location is one of the most important points in appendectomies.

2.6 Large Intestine

The large intestine has six portions: cecum, ascending colon, transverse colon, descending colon, sigmoid colon, and rectum. The cecum and appendix were described in the previous section. Next, we will address the anatomical aspects of other segments of the large intestine. The ascending colon begins above the ileocecal valve and extends to the region of the right colic (hepatic) flexure. The transverse colon is located between the two colic flexures, right and left. The descending colon extends from the left (splenic) colic flexure to the upper narrow of the pelvis; the sigmoid colon extends from the upper narrow of the pelvis to the region of the second or third sacral vertebra, where the rectum begins. There are spaces between the ascending and descending colon and the abdominal wall, called paracolic gutters (Williams et al. 1995). These gutters are important in the dissemination of collections within the abdominal cavity, constituting the place where these collections usually lodge when the individual assumes the orthostatic position (Fig. 2.6). The colic flexures are attached to the diaphragm by the phrenicocolic ligaments. These ligaments, during a colectomy, need to be sectioned in order for the colon to be mobilized. The spleen rests on the left phrenicocolic ligament, an anatomical relationship of importance during splenectomy and left colectomy surgeries. The ascending and descending colon are partially covered by the peritoneum, showing little mobility, while the transverse and sigmoid colons have long peritoneal folds (mesos), being quite motile. The sigmoid colon is the region with the greatest mobility of the colon and may present anomalies in the implantation of its meso, which can lead to the occurrence of sigmoid torsion, the volvulus, a condition that can evolve with obstruction and necrosis of the loop (Fig. 2.7).

The arteries that supply the large intestine are the superior mesenteric (irrigates the portion of the large intestine corresponding to the middle intestine of the embryo) and the inferior mesenteric (irrigates the portion of the large intestine corresponding to the posterior intestine). The region irrigated by the superior mesenteric artery can

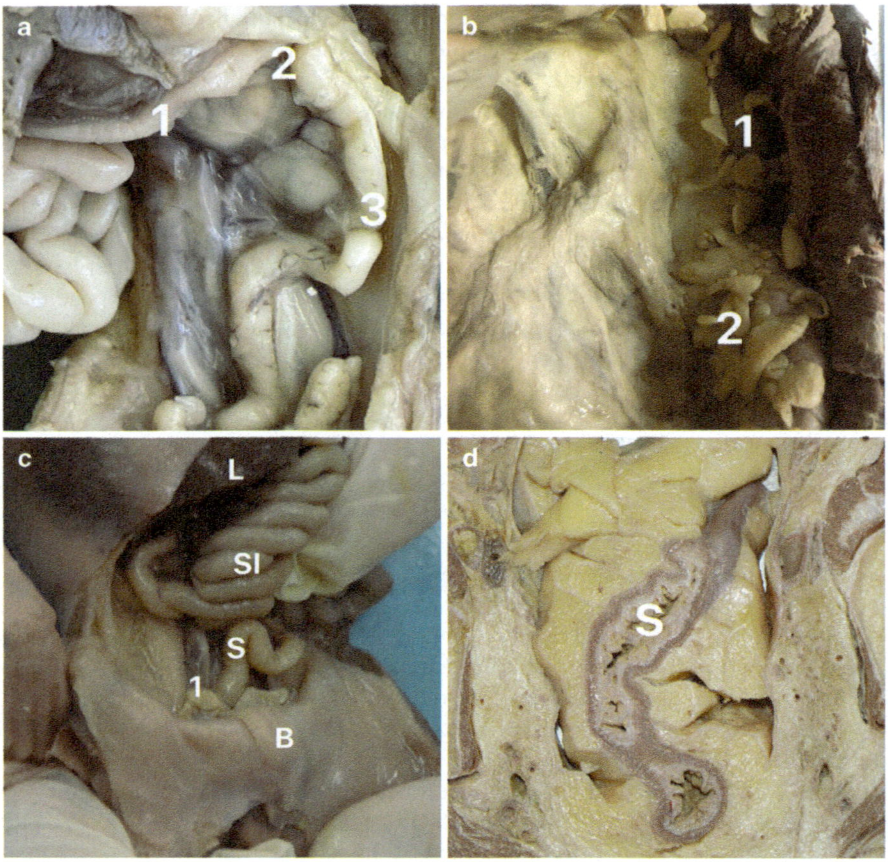

Fig. 2.7 Left colon. (**a**) The figure shows the abdominal cavity of a human male fetus of the second gestational trimester; we can observe the transverse colon (1), the splenic flexure (2), and the descending colon (3). (**b**) The figure shows the abdominal cavity of a formalized corpus; we can observe the relationships between the descending colon (1) and the sigmoid colon (2). (**c**) The figure shows the abdominal cavity of a human female fetus of the second gestational trimester; we can observe the small intestine (SI), the sigmoid colon (S), the ovarium (1), the liver (L), and the bladder (B). (**d**) The figure shows an inferior view of a transverse section at the level of the fifth lumbar vertebra in a frozen corpus. We can observe the relationships of the sigmoid colon (S)

also be called the right colon, while the portion irrigated by the inferior mesenteric artery can be called the left colon. The superior mesenteric artery originates three branches: ileocolic artery (already described above), right colic artery, and middle colic artery. The inferior mesenteric artery originates the left colic artery, the sigmoid arteries, and the superior rectal artery (Netter 1978). The right colic artery mainly irrigates the ascending colon. The middle colic artery is responsible for irrigating most of the transverse colon, while the left colic artery irrigates a small portion of the transverse colon and the descending colon. The sigmoid arteries supply the sigmoid, and the rectal artery, the upper part of the rectum. Colic arteries anastomose widely, forming a large vascular arch along the wall of the loops (arch of Riolan or mesenteric meandering artery). The anastomotic arch along the transverse colon was formerly called the marginal artery of Drummond. Between the sigmoid arteries and the superior rectal artery, there is an area in the transition between the sigmoid colon and the rectum that is poor in the arches, presenting a deficient collateral circulation. This area is considered a critical point during resections of this segment of the intestine and is called the Sudeck critical point.

2.7 Spleen

The spleen is an organ derived from the primitive intestine and vascularized by intestinal vessels, but it has no action on the digestive system, and its function is related to the lymphoid system. It is a flabby, vascular organ and usually aubergine in color. It is totally surrounded by the peritoneum, except in the splenic hilum, where it is related to the tail of the pancreas, and the splenic arteries and veins also pass (Williams et al. 1995). The spleen is located on the upper floor of the abdomen (upper left quadrant), in the left subphrenic site, above the transverse mesocolon and left colic flexure (splenic), at the level of the 9th, 10th, and 11th ribs on the same side, with its axis located at height and along the 10th rib (a fact that assumes particular importance in blunt abdominal trauma). It is separated from the ribs by the diaphragm muscle and the costodiaphragmatic recess (Fig. 2.8).

Under normal conditions, the spleen has three faces, three margins, and two extremities. Faces: (1) diaphragmatic (lateral)—it is extensive, convex, and shaped over the concavity of the diaphragmatic dome; (2) gastric (anteromedial)—it is concave, close to the convexity of the greater curvature of the stomach; in the posterior part of this surface is the splenic hilum, a depression full of vascular impressions orientated according to the bundle of the organ, depressions with irregular contours of very variable height; and (3) renal (posteromedial)—it is concave, also orientated downward inferiorly; lays on to the kidney and the left adrenal gland.

Margins separate the faces: (1) upper (anterior)—it is convex and irregular, marked by two or three very characteristic notches that persist in the spleens with splenomegaly, which allows recognizing the splenic quality of some abdominal tumors; (2) inferior (posteroinferior)—it is rounded and leans against the upper pole of the left kidney; and (3) internal (medial)—it is superiorly thin, inferiorly enlarged; it forks out a colic face, which is the colic impression. Poles are extremities: (1) upper pole—it is posterior, rounded, and slightly depressed medially; and (2) anterior pole—inferior and very striking, sometimes when the colic impression is wide, it becomes rounded. Anatomical relationships: The diaphragmatic surface is related to the costal part of the diaphragm. The visceral surface presents the surfaces: gastric, renal, and colic (Bergman et al. 1988; Williams et al. 1995). The gastric surface is related to the stomach. A long fissure, the hilum, present in the

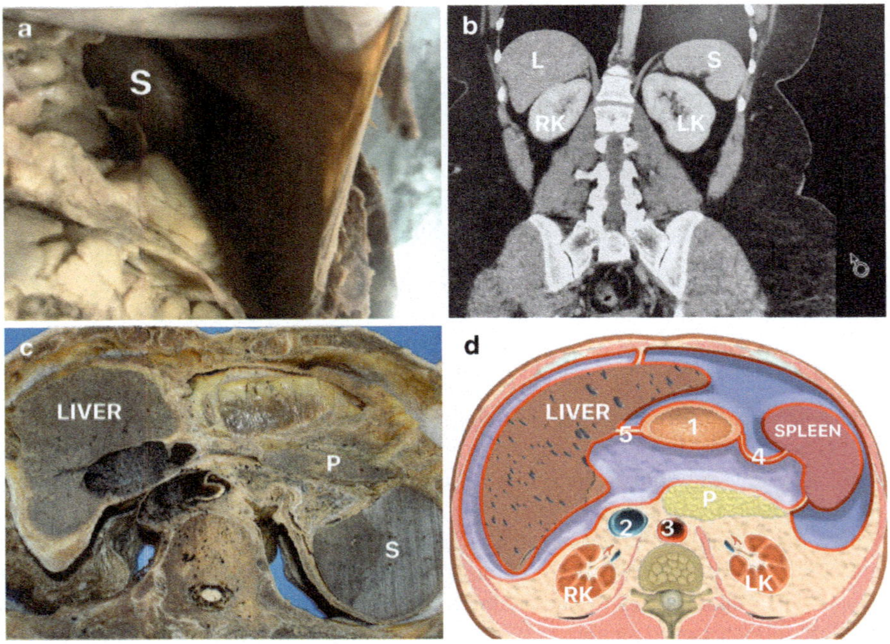

Fig. 2.8 Spleen relationships. (**a**) The figure shows the abdominal cavity – the supramesocolic region of a formalized corpus; we can observe the relationships of the spleen (S). (**b**) The figure shows an abdominal computerized tomography in a frontal section showing the relationships of the spleen (S). RK, right kidney; LK, left kidney; L, liver. (**c**) The figure shows an inferior view of a transverse section at the level of the 12th thoracic vertebra in a frozen corpus. We can observe the relationships of the spleen (S) with the pancreas (P). (**d**) Schematic drawing of an inferior view of a transverse section at the level of the first lumbar vertebra; we can observe the relationships between the stomach (1) with spleen and the liver; RK, right kidney; LK, left kidney; 2, inferior vena cava; 3, aorta artery; 4, gastrosplenic ligament; 5, hepatogastric ligament

lower part of the spleen, is perforated by vessels and nerves. The renal surface of the upper part of the visceral surface is related to the left kidney and sometimes to the left adrenal gland. The colic surface, at the anterior end, is related to the left colic flexure. The tail of the pancreas can reach the spleen, between the colic surface and the hilum (Fig. 2.9). The spleen is completely surrounded by the peritoneum, except in the splenic hilum. A peritoneal reflection can start from the lower pole of the greater omentum, and it may rupture during surgical retraction of the stomach to the right, resulting in bleeding. The spleen develops in the dorsal mesogastrium and remains connected to the stomach by the gastrosplenic (gastrolienal) ligament, and to the body wall and kidney, by the phrenicosplenic (phrenicolienal) ligament, the lower part of which is called the splenorenal ligament (lienorenal). The splenorenal ligament passes through the splenic vessels and contains the tail of

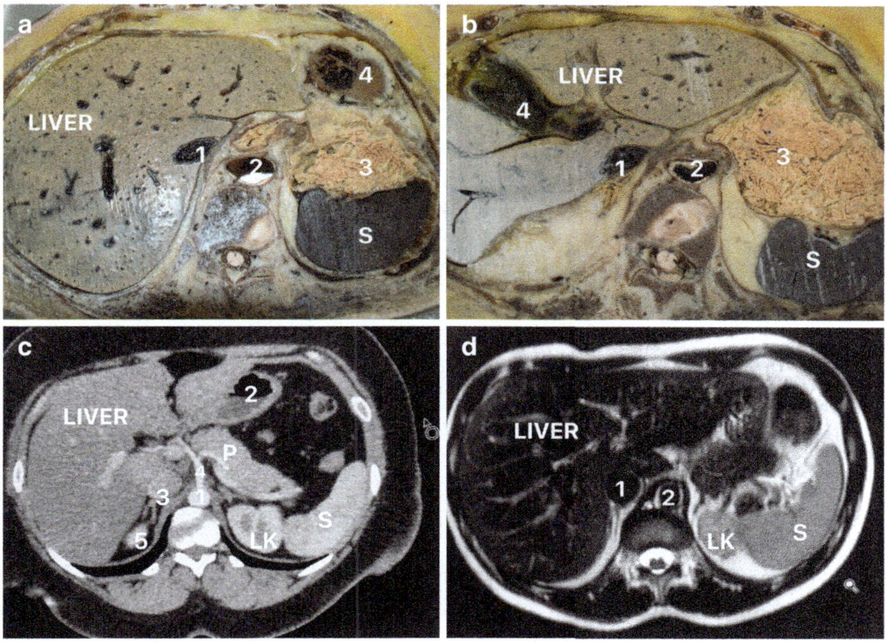

Fig. 2.9 Spleen relationships. (**a**) The figure shows an inferior view of a transverse section at the level of the seventh thoracic vertebra in a frozen corpus. We can observe the relationships of the spleen (S) with the stomach (3); 1, inferior vena cava; 2, aorta artery; 4, lower portion of the left ventriculum. (**b**) The figure shows an inferior view of a transverse section at the level of the tenth thoracic vertebra in a frozen corpus. We can observe the relationships of the spleen (S) with the stomach (3) in this lower section; 1, inferior vena cava; 2, abdominal aorta artery; 4, gallbladder. (**c**) The figure shows an abdominal computerized tomography at the level of the 12th thoracic vertebra; we can observe the relationship of the spleen (S) with the pancreas (P) and the left kidney (LK). 1, abdominal aorta artery; 2, stomach; 3, inferior vena cava; 4, superior mesenteric artery; 5, right kidney. (**d**) The figure shows an abdominal magnetic resonance image (MRI) at the level of the 12th thoracic vertebra; we can observe the relationship of the spleen (S) with the left kidney (LK). 1, inferior vena cava; 2, abdominal aorta artery

the pancreas (Williams et al. 1995). The arterial supply comes from the splenic artery, which is the largest branch of the celiac trunk. It has a tortuous path along the upper margin of the pancreas. Then, it is divided into approximately six branches, which penetrate the spleen through the hilum. The common variations of this arrangement are shown in Fig. 2.9.

The splenic vein leaves the hilum and subsequently transits to the tail and body of the pancreas. Behind the pancreatic neck, the splenic vein joins the superior mesenteric vein to form the portal vein of the liver.

2.8 Pancreas

The pancreas is located in the upper and posterior 2/3 of the abdominal cavity, hidden by organs of the epigastric region. On the surface, it is delimited in the upper left quadrant; however, it is not accessible to physical examination. It is located retroperitoneally and transversely, from right to left, and from bottom to top, molded to the posterior abdominal wall, behind the stomach between the duodenum, on the right, and the spleen, on the left, and, for the most part, behind of the lesser omentum. The transverse mesocolon is attached to its anterior margin. The pancreas extends from the right side of the L1, L2, and sometimes L3 vertebrae to the left, up to the spleen hilum, passing in front of the inferior vena cava and the aortic artery. Thus, it projects a sinuous profile in a cross section, which is not evident when viewed from the front (Williams et al. 1995).

This organ consists of the head, neck, body, and tail. Its shape is elongated and lobulated (anfractuous) due to its surface characteristics, as well as its structure, presenting an analogy with the salivary glands. The head, wide and flat, adapts perfectly to the concave curvature of the duodenum, resembling the shape of a disc that has a small right anterior orientation. To the left, the head remains in the lap (isthmus); from the bottom of the head to the left, the uncinate process is projected. The neck is compressed by the superior mesenteric vessels (artery and vein) that are located in the pancreatic notch, a deep groove on the posterior face of the neck. The uncinate process is located below the neck of the pancreas and behind the superior mesenteric vessels. The body is located above the duodenojejunal flexure and the left kidney, where it tapers in the tail, which is often understood by the splenorenal ligament.

2.8.1 Anatomical Relationships

The anterior and posterior sides of the pancreas relate to a large number of organs. The base of the transverse mesocolon attaches to the head and along the lower border of the neck and body. Consequently, most of the anterior face turns toward the omental sac and is related, through the sac, to the stomach. Above the lesser curvature, the lesser omentum, and through it, the liver can also be put in contact with the pancreas. The first part of the duodenum is above or on the anterior surface of the head. The lower part of the head and the narrow underside of the neck and body face the infracolic compartment and come into contact with bowel loops (Netter 1978). The posterior aspect of the head of the pancreas is located between the hilum of the right kidney and its vessels, over the portal vein and over the inferior vena cava. Next to the duodenum, the upper part of the posterior face of the head is crossed by the bile duct. The neck and uncinate processes are located in front of the aortic artery. The body crosses the left kidney, just above the hilum, also relating to the left adrenal gland and the right pillar of the diaphragm muscle (Fig. 2.10).

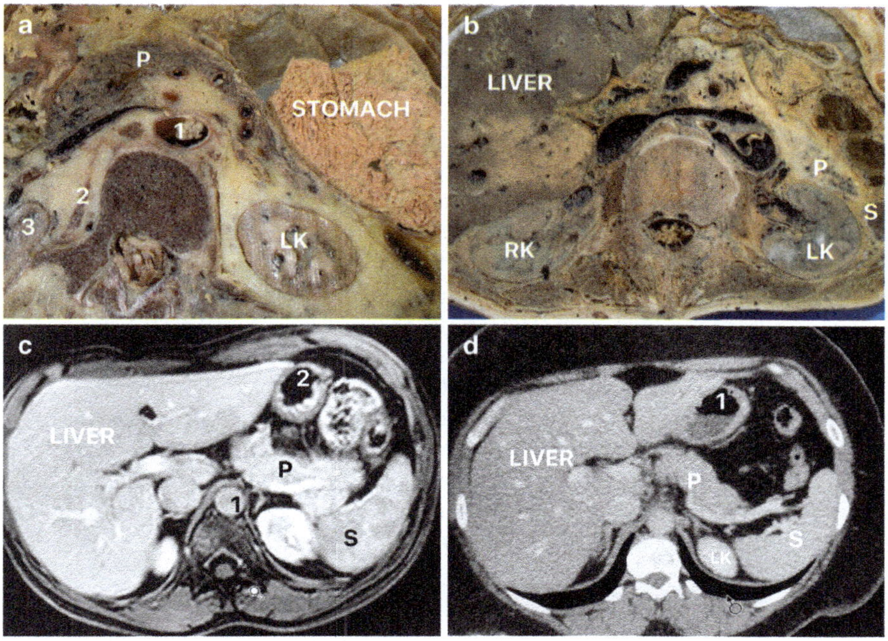

Fig. 2.10 Pancreas. (**a**) The figure shows an inferior view of a transverse section at the level of the 12th thoracic vertebra in a frozen corpus. We can observe the relationship between the pancreas (P) and the stomach; LK, left kidney; 1, abdominal aorta artery; 2, right diaphragmatic pillar; 3, right kidney. (**b**) The figure shows an inferior view of a transverse section at the level of the second lumbar vertebra in a frozen corpus. We can observe the relationship between the pancreas (P) with the spleen (S) and the left kidney (LK). RK, right kidney. (**c**) The figure shows an abdominal magnetic resonance image (MRI) at the level of the 12th thoracic vertebra; we can observe the relationship of the pancreas (P) with the spleen (S). 1, abdominal aorta artery; 2, stomach. (**d**) The figure shows an abdominal computerized tomography at the level of the 11th thoracic vertebra; we can observe the relationship of the spleen (S) with the pancreas (P) and the left kidney (LK). 1, stomach

In addition to the superior mesenteric vessels, the celiac trunk and two of its branches are closely related to the pancreas. The celiac trunk originates at the level of the upper margin of the pancreas and can often be embedded in pancreatic tissue; the common hepatic artery runs to the right along the upper border of the neck and head; the splenic artery runs to the left along the upper margin of the body and crosses the tail onward. The splenic vein is behind the pancreas and is reached there by the inferior mesenteric vein. The confluence of the splenic and superior mesenteric veins forms the hepatic portal vein on the posterior face of the pancreas. The tip of the tail has contact with the spleen's hilum, and in many cases, it should not be damaged during splenectomy.

Pancreatic Duct: The main pancreatic duct begins at the tail and passes through the entire length of the pancreas, receiving numerous tributary ducts along the way. It opens, with the bile duct, in the second part of the duodenum, approximately in its half, in the greater papilla of the duodenum (Aljiffry et al. 2020). The accessory pancreatic duct (or duct of Santorini), when present, drains the upper part of the head and then opens into the duodenum a short distance above the main duct in the minor papilla of the duodenum. The accessory pancreatic duct often communicates with the main duct (Zhou et al. 2006). The choledochal duct is not a pancreatic duct but has an intrapancreatic segment. Halfway through a third part of its path, it is located in a groove on the posterior face of the pancreas so that the choledochal duct comes into contact with the main pancreatic duct. The common bile duct crosses the medial wall of the second portion of the duodenum. It usually joins with the pancreatic duct and, together, they open in a small dilation, the hepatopancreatic ampulla (or Vater papilla). The ampulla opens in the lumen of the duodenum through the major papilla of the duodenum. The terminal parts of both ducts and the ampulla are surrounded by circular muscular extracts known as the sphincter muscle of the hepatopancreatic ampulla (sphincter of Oddi). Occasionally, the choledochal cyst and pancreatic ducts open separately in the duodenum.

Variations in biliary anatomy and pancreatic duct are very diverse and extend from the intrahepatic biliary system down to the pancreas, and the knowledge of the possible variations could provide the surgeon with better control to prevent any unwanted outcomes (Aljiffry et al. 2020).

2.8.2 Vessels

The duodenum and the pancreas are irrigated by the branches of the celiac trunk and the superior mesenteric artery. The first portion of the duodenum has a poor blood supply provided by small branches of the gastroduodenal artery. The blood supply of the second, third, and fourth portions of the duodenum and the head of the pancreas are provided by two parallel arterial arches, located inside the duodenal concavity, on the surface or very close with the head of the pancreas. The arches are fed, superiorly, by branches of the gastroduodenal artery and inferiorly, by branches of the superior mesenteric artery. The neck, body, and tail of the pancreas are irrigated by the splenic artery (Williams et al. 1995). The first portion of the duodenum is supplied by the supraduodenal and retroduodenal arteries, branches of the gastroduodenal. These vessels can be emitted by the gastroduodenal artery directly or by any of its branches. The artery arches in the head of the pancreas are made up of four pancreatic-duodenal arteries that freely anastomose, anterior and posterior vessels, at the top, emitted by the gastroduodenal artery and anterior and posterior vessels, at the bottom, provided by the superior mesenteric artery. Now the anterior arch, now the posterior, may be incomplete, but, in relation to such small vessels, its constancy is noteworthy. The anterior superior pancreaticoduodenal artery represents one of the terminal branches of the gastroduodenal artery that descends behind the first portion of the duodenum. The other terminal branch, the right gastro-omental artery (or gastroepiploic artery), has already been demarcated in the greater gastric curvature of the stomach. The posterior superior pancreaticoduodenal artery is supplied by the gastroduodenal artery before dividing into its terminal branches. The posterior superior pancreaticoduodenal artery occurs inferiorly along the left side of the choledochal cyst duct. The two pancreaticoduodenal arteries come from a common trunk (the first branch emitted by the superior mesenteric artery) that divides into the anterior and posterior branches. These vessels ascend very close with the head of the pancreas or on its surface to anastomose with the respective arteries derived from the gastroduodenal. Both the anterior and posterior arches provide branches to the head of the pancreas and a series of straight vessels for the second, third, and fourth portions of the duodenum. The blood supply of the fourth portion and the duodenojejunal flexure is increased by the duodenal branches of the superior mesenteric artery and the first jejunal artery.

2.9 Liver

The liver is the largest gland in the body, weighs approximately 1500 grams, and corresponds to 2% of the bodyweight of an adult and 5% of the weight of a newborn due to liver hematopoiesis at this stage of life. It is located in the right hypochondrium, under the diaphragm, and protected by the ribs and lower costal cartilages. It extends from the fourth intercostal space to the lower edge of the costal fence, and its size can be assessed clinically by liver percussion, which is approximately 10 cm in the midclavicular line. Liver faces: The liver has two faces, a convex and upper face, called the diaphragmatic face and another lower and posterior face, which is in contact with the abdominal viscera, called the visceral face. The diaphragmatic face is covered by the peritoneum except in the bare area of the liver, which is in direct contact with the diaphragm. The bare area is bounded by reflections from the peritoneum to the diaphragm through the anterior and posterior laminae of the coronary ligament. The union of the blades, laterally, forms the right triangular ligament. The left triangular ligament is to the left of the sickle cell ligament. The liver is separated from the diaphragm by the right and left subphrenic recesses on either side of the sickle cell ligament. There is a communication between the right subphrenic recess and the hepatorenal or Morison recess (between the liver and the right kidney). The sickle cell ligament is composed of two peritoneal laminae that separate on the diaphragmatic face: the left lamina is continuous with the left triangular ligament and the right lamina with the anterior lamina of the coronary ligament (Netter 1978; Williams et al. 1995) (Fig. 2.11).

The visceral face is also lined by the peritoneum, except at the liver port, where the vessels and ducts enter and leave the liver, and the bed of the gallbladder. Some organs contact this face, for example, the right kidney, the right adrenal gland, the right colic flexure and the transverse colon, the abdominal part of the esophagus, the stomach, the duodenum, and the gallbladder. The presence of fissures and grooves on the visceral face forms the image of the letter H. The crossbar of the H is the portal of the liver, where the hepatic hilum formed by the portal vein, hepatic artery, and bile ducts is located. The right portion of the H is formed anteriorly by the gallbladder fossa and posteriorly by the groove of the inferior vena cava. On the left, H

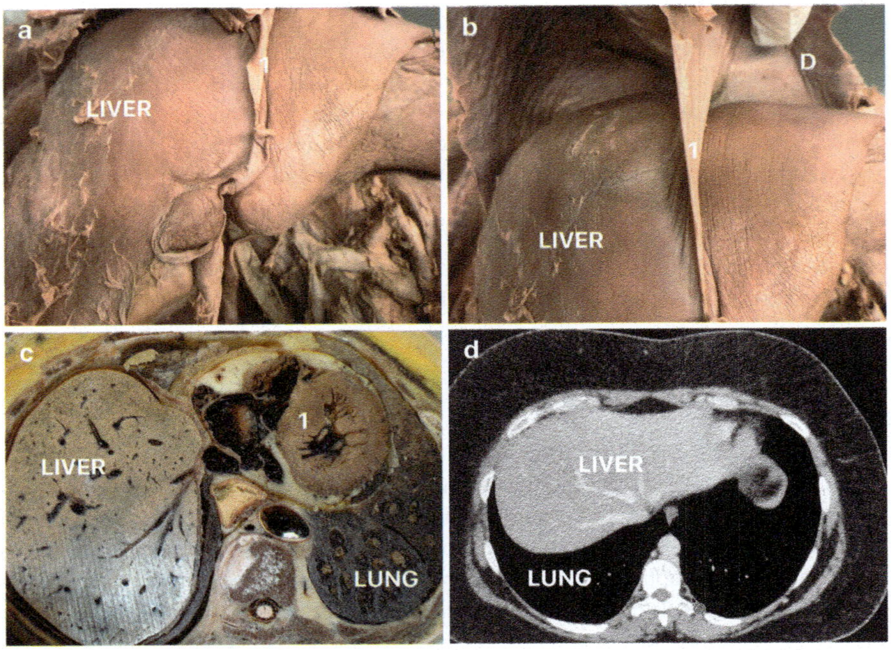

Fig. 2.11 Liver. (**a**) The figure shows the abdominal cavity of a formalized corpus; we can observe the liver and the falciform ligament (1). (**b**) The figure shows the abdominal cavity of the same formalized corpus; we can observe the relationships between the liver with the diaphragm muscle (D) and the falciform ligament (1). (**c**) The figure shows an inferior view of a transverse section at the level of the sixth thoracic vertebra in a frozen corpus. We can observe the relationship between the liver with the lung, and we also observe the left ventriculum (1). (**d**) The figure shows an abdominal computerized tomography at the level of the eighth thoracic vertebra; we can observe the relationship between the liver and the right lung

is formed anteriorly by the fissure of the round ligament of the liver and subsequently by the fissure of the venous ligament. The round ligament is formed by the obliterated umbilical vein, and the venous ligament, by fibrosis of the venous duct that has no function after birth. The presence of hepatic H forms two smaller lobes on the visceral surface: the square lobe, anteriorly, and the caudate lobe, posteriorly. The fissure of the round ligament and the fossa of the round ligament delimit the square lobe, and the fissure of the venous ligament and the groove of the inferior vena cava separate the caudate lobe.

The liver port is a transverse and approximately 5 cm deep fissure that separates the square lobe from the caudate lobe. It enters the portal vein, the hepatic artery, and the hepatic nerve plexus, as well as exiting through the right and left hepatic ducts and lymphatic vessels. All these structures are called hepatic pedicles. The anatomical relationship within the pedicle is as follows: the hepatic duct is to the right of the hepatic artery and the portal vein is posterior to the hepatic duct and the hepatic artery. The peritoneum that surrounds these structures forms the minor omentum that joins the stomach and the first portion of the duodenum to the liver through the hepatogastric and hepatoduodenal ligaments. The lesser omentum is the anterior limit of the epiploic foramen (Winslow), which communicates the peritoneal cavity with the omental sac (Williams et al. 1995). Below we can see a series of transverse cuts showing the liver relations (Fig. 2.12).

Segmental anatomy: Externally, the liver is divided into four lobes called the right, left, square, and caudate lobe. The sickle cell ligament separates the right and left lobes on the diaphragmatic side, and the square and caudate lobes are formed by the hepatic H on the underside of the liver. Current concepts about hepatic segmentation differ from the external division due to the reflections of the peritoneum, mainly due to the sickle cell ligament. Segmentation does not use the hepatic surface as a reference but rather the distribution of the main branches of the portal vein, hepatic artery, or intrahepatic bile ducts. These three structures form the hepatic triad that branches internally, dividing the right and left lobes. The division is

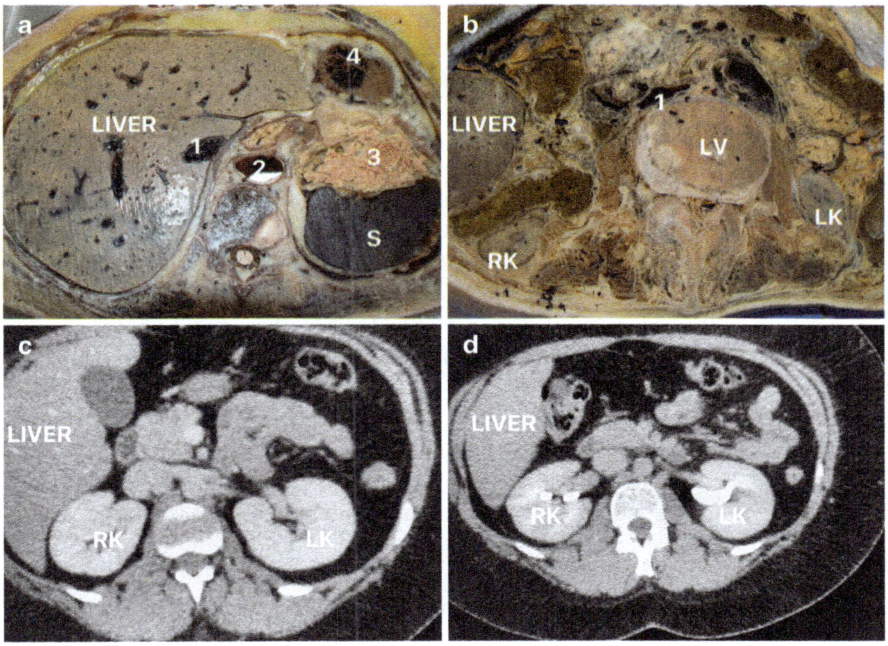

Fig. 2.12 Liver. (**a**) The figure shows an inferior view of a transverse section at the level of the ninth thoracic vertebra in a frozen corpus. We can observe the liver, the spleen (S), and the stomach (3). 1, inferior vena cava; 2, aorta artery. (**b**) The figure shows an inferior view of a transverse section at the level of the second lumbar vertebra (LV) in a frozen corpus. We can observe the relationship between the liver and the right kidney (RK); LK, left kidney; 1, inferior vena cava. (**c**) The figure shows an abdominal computerized tomography at the level of the 12th vertebra; we can observe the relationship between the liver and the right kidney (RK). LK, left kidney. (**d**) The figure shows an abdominal computerized tomography at the level of the second thoracic vertebra; we can observe the relationship between the liver and the right kidney (RK). LK, left kidney

followed from the hilum to the hepatic parenchyma by the perivascular fibrous capsule (Glisson's capsule). By the internal division, the right lobe is smaller than that previously described by the sickle cell ligament and has its limit at the level of the gallbladder fossa and the groove of the inferior vena cava. On the other hand, the left lobe is larger than that suggested externally and includes the square and caudate lobes, which are to the right of the sickle cell ligament. Currently, there are two classifications for the division of liver segments: the so-called American system and the other is the French segmental system. Under the American system, the left lobe of the liver is divided into a medial segment, which is to the right of the sickle cell ligament, and a lateral segment, which is to the left of the same ligament. The right lobe has an anterior and a posterior segment. This division into four segments is the basis for the French segmental system that serves as the basis for liver resections.

In the French system (Soupault and Couinaud), the liver is divided, taking into account the hepatic venous drainage, the portal, arterial, and biliary anatomy; in this way, eight segments are formed, four on the right side and three on the left in addition to the caudate lobe. Segment I corresponds to the caudate lobe, II to IV to the left lobe, and segments V to VIII to the right lobe. When comparing the French system with the American system, we find that segments II and III of the French system together form the American side segment. In the right lobe, segments V and VIII are the anterior American segment and segments VI and VII form the posterior American segment. Segment I receives irrigation from the right and left branches of the portal vein and the hepatic artery, and venous drainage is performed directly into the inferior vena cava. For this reason, segment I is autonomous. Segmentation based on the portal triad and hepatic venous drainage is important, since a ligation of the venous branches, even with the maintenance of the flow through the branches of the hepatic artery and the portal vein, can cause necrosis of the hepatic parenchyma. The liver receives a volume of blood corresponding to 25% of cardiac output per minute in an adult at rest. Of this volume, approximately 75% comes from the portal vein and the remainder from the hepatic artery itself. The port system drains one capillary system to another capillary system; that is, it joins two capillary systems.

The portal vein is formed by the union of the superior mesenteric vein with the splenic vein, posterior to the pancreas neck, and anteriorly to the inferior vena cava, at the level of the second lumbar vertebra. In 25% of cases, the inferior mesenteric vein joins the splenic and superior mesenteric veins to form the portal vein. The portal vein has a diameter ranging from 1 to 3 cm and between 5 and 8 cm in length. After its formation, it subsequently ascends to the duodenum and pancreas to form part of the hepatoduodenal ligament. Within the hepatic pedicle, the portal vein is posterior to the hepatic artery and bile duct. At the hepatic hilar region, it divides into right and left branches. Within the hepatic parenchyma, the right branch forms the anterior and posterior branches for drainage of the right lobe of the liver. The left branch has a transverse path through the liver port and irrigates the lower part of the lateral segment of the left lobe and the medial segment. The vertical portion of the branch is anastomosed with the umbilical vein reopened within the sickle cell ligament, forming a collateral circulation when there is portal hypertension. The veins that supply the caudate lobe are formed from the left branch.

The common hepatic artery originates from the celiac trunk, inside the omental sac, passes through the upper margin of the pancreas and posteriorly to the duodenum, where the gastroduodenal and right gastric arteries are born. From this point on, the artery is called proper hepatic artery and ascends through the hepatoduodenal ligament following the choledochal duct on the left and anteriorly to the portal vein. Before penetrating the liver, the proper hepatic artery divides into the right and left hepatic arteries. The right hepatic artery is the larger of the two and emits the cystic artery for irrigation of the gallbladder.

Variants in arterial anatomy of the celiac trunk and hepatic artery are common findings, and this knowledge is very important in liver transplantation, laparoscopic procedures, and minimally invasive radiological abdominal interventions (Arifuzzaman et al. 2017).

In cases of trauma with extensive liver damage, a manual compression maneuver of the hepatic pedicle can be performed (Pringle maneuver), which consists of the occlusion of the hepatic artery and the portal vein, when these vessels pass through the hepatoduodenal ligament. The surgeon inserts his index finger into the epiploic foramen and, with the thumb placed anteriorly to the hepatoduodenal ligament, compresses the hepatic pedicle, which leads to a temporary decrease in bleeding. Hepatic veins: The venous drainage of the liver begins with the drainage of the hepatic sinusoids that are divided into lobules. The irrigation of the sinusoids is done through terminal branches of the hepatic artery and the portal vein and drainage through the central vein. The central veins join and form the sublobular veins that end in the collecting veins. The hepatic veins are formed from the collecting veins. There are three hepatic veins: right, left, and middle. The short extrahepatic path of these veins makes surgical access difficult to control bleeding. Another cause of death in lesions of these veins is the gas embolism that can occur during laparotomy. The right hepatic vein is the largest of these three and drains the right lobe from the liver. It is located between the anterior and posterior segments (American system) and drains these two segments and flows separately into the inferior vena cava. The left hepatic vein drains the lateral segment of the left lobe, and the middle hepatic vein drains the medial segment. The left and middle hepatic veins flow together into the inferior vena cava in 80% of cases. The caudate lobe drains separately into the inferior vena cava. In thrombosis of the main hepatic vein (Budd-Chiari syndrome), this autonomous drainage becomes important, which is why there is hypertrophy of the caudate lobe.

References

Aljiffry M, Abbas M, Wazzan MAM, Abduljabbar AH, Aloufi S, Aljahdli E. Biliary anatomy and pancreatic duct variations: a cross-sectional study. Saudi J Gastroenterol. 2020;26:188–94.

Arifuzzaman M, Sawaira S, Naqvi N, Adel H, Syed, Adil SO, Rasool M, Hussain M. Anatomical variants of celiac trunk, hepatic and renal arteries in a population of developing country using multidetector computed tomography angiography. J Ayub Med Coll Abbottabad. 2017;29(3):450–4.

Bergman RA, Thompson SA, Afifi AK, Saadeh FA. Compendium of human anatomic variation: catalog, atlas and world literature. Baltimore: Urban & Schwarzenberg; 1988.

Hollishead WH, Rosse C. Anatomia. 4th ed. Rio de Janeiro: Interlivros; 1991.

Latarjet M, Ruiz-Liard A. Anatomia Humana, vol. 2. 2nd. ed., capítulo 20. São Paulo: Panamericana; 1993. p. 1325.

Moore KL, Dalley AF. Anatomia Orientada para Clínica. 4th ed., capítulo 2, Rio de Janeiro: Guanabara Koogan; 2001. p. 127–48.

Netter F. Reproductive system, vol. 2. New Jersey: Ciba; 1978.

Testut L, Jacob O. Anatomia Topográfica. Tomo I, Barcelona, Salvat; 1926. p. 72–287.

Williams PL, Bannister LH, Berry MM, Dyson M, Dussek JE, Ferguson MWJ. Gray's anatomy, vol. 2. 38th ed. Edinburgh: Churchill-Livingstone; 1995. p. 974–1029.

Xu H, Xiang K, He M, Tian F, Wu Y. Visualization of the omental bursa and its spatial relationships to left subphrenic extraperitoneal spaces by the second Chinese Visible Human model. Surg Radiol Anat. 2015;37:473–81.

Yassi R, Cheng LK, Al-Ali S, Sands G, Gerneke D, Legrice I, Pullan AJ, Windsor JA. Three-dimensional high-resolution reconstruction of the human gastro-oesophageal junction. Clin Anat. 2010;23(3):287–96.

Zhou ZM, Fang CH, Huang LW, Zhong SZ, Wang BL, Zhou WY. Three dimensional reconstruction of the pancreas based on the virtual Chinese human—female number 1. Postgrad Med J. 2006;82:392–6.

Chapter 3
Sectional Anatomy of the Retroperitoneum

Luciano Alves Favorito, Natasha T. Logsdon, and Francisco J. B. Sampaio

3.1 Retroperitoneum

The retroperitoneum is located behind the abdominal cavity. The retroperitoneal space is a three-dimensional space, ranging from the diaphragm to the pelvic extraperitoneal space. In this region, there is a large amount of adipose tissue and some organs of great importance, such as the kidney, the adrenal glands, the pancreas, the ascending and descending colon, and the large abdominal vessels (Sampaio 1996).

L. A. Favorito (✉)
Urogenital Research Unit, Rio de Janeiro State University, Rio de Janeiro, RJ, Brazil

IDOMED-UNESA, Rio de Janeiro, RJ, Brazil

The National Council for Scientific and Technological Development (CNPq), Lago Sul, Federal District, Brazil

The Rio de Janeiro State Research Foundation (FAPERJ), Rio de Janeiro, RJ, Brazil

N. T. Logsdon
University Center Geraldo di Biasi, Rio de Janeiro, RJ, Brazil

Urogenital Research Unit, Rio de Janeiro State University, Rio de Janeiro, RJ, Brazil

F. J. B. Sampaio
Urogenital Research Unit, Rio de Janeiro State University, Rio de Janeiro, RJ, Brazil

The National Council for Scientific and Technological Development (CNPq), Lago Sul, Federal District, Brazil

The Rio de Janeiro State Research Foundation (FAPERJ), Rio de Janeiro, RJ, Brazil
e-mail: sampaio@uerj.br

L. A. Favorito, N. T. Logsdon (eds.), *Atlas of Sectional Anatomy*,
https://doi.org/10.1007/978-3-030-91688-6_3

3.1.1 Kidneys

Kidneys are paired organs located at the retroperitoneum close to the posterior wall of the abdomen. They have a unique shape, with an upper pole (superior end) and a lower pole (inferior end), with a convex lateral border and a median concave border. The median border presents a depression, the kidney hilum, with the renal vessels, and renal pelvis or basinet (Williams et al. 1995) (Fig. 3.1).

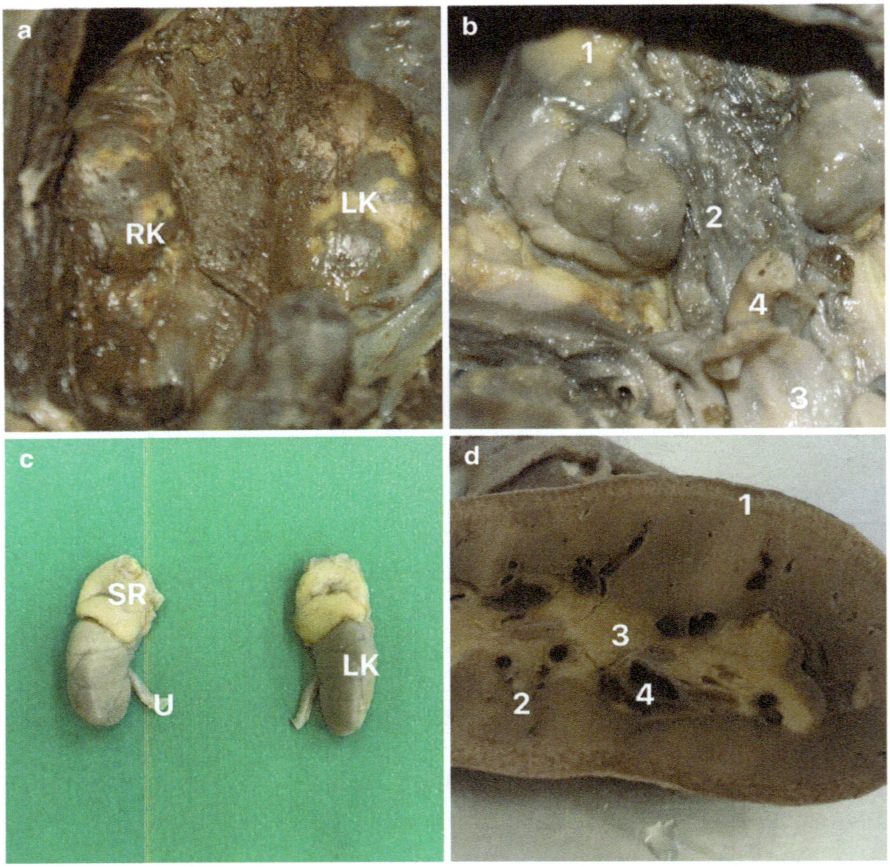

Fig. 3.1 (**a**) Fetus with 18 weeks post-conception (WPC); we can observe the retroperitoneum after the abdominal organs were removed and the kidney position. RK, right kidney; LK, left kidney. (**b**) In the same fetus after the fascias and fat dissection, we can observe the relationship of the right kidney with the adrenal gland (1) and with the inferior vena cava (2); we can also observe the bladder (3) and the urachus (4) in this fetus. (**c**) The kidneys were removed from the retroperitoneum in the fetus with 18 WPC showing the relationships of the kidney upper pole with the adrenal gland (SR) and the kidney lower pole with the ureter (u). (**d**) The figure shows a frontal section of an adult kidney with the renal cortex (1), renal medulla (2), renal pelvis (3), and renal vessels (4)

The kidneys are located at the posterior abdominal wall, in contact with the major psoas muscle in each side, and therefore their longitudinal axis parallels the oblique direction of the psoas. The major psoas is cone-shaped; therefore, the kidneys are also bent posteriorly to the longitudinal axis, and the upper poles are more medially and inferiorly located than the lower poles. The renal hilum is anteriorly rotated over the psoas, positioning the kidney borders dorsally (Fig. 3.2). Therefore, the kidneys are angulated at 30–50° posteriorly to the frontal plane (Williams et al. 1995).

The surface of the kidneys is covered by a fibrous tissue, called renal capsule (real renal capsule). Each kidney is involved by a mass of adipose tissue (perirenal fat) located between the peritoneum and the posterior abdominal wall (Fig. 3.1). The perirenal fat is enveloped by the renal fascia, a membranous structure consisting of fibrous connective tissue covering the kidney, perirenal fat, the ipsilateral ureter, the adrenal gland, and gonadal vessels from the anterior and posterior aspects in the retroperitoneum (Fig. 3.3). The perirenal fat belongs to the connective tissue of the gonad and the adrenal gland, and the ureter, adrenal, and kidneys are separated from perirenal fat, and it is located on the dorsal side inside the renal fascia (Ochi et al. 2020) (Fig. 3.3).

The renal fascia has two parts: the anterior lamina (Gerota's fascia) and the posterior lamina (Zuckerkandl's fascia or fascia retrorenalis) (Ochi et al. 2020). Since the advent of computerized tomography (CT), the concept of "interfascial plane" has been reported, with the renal fascia having been recognized as a multilayered membranous structure with a potential inner space rather than as a single membrane (MindelI et al. 1995; Molmenti et al. 1996). The renal fascia is circled anterior and posteriorly by another layer of fat tissue, with various thicknesses, called para-renal

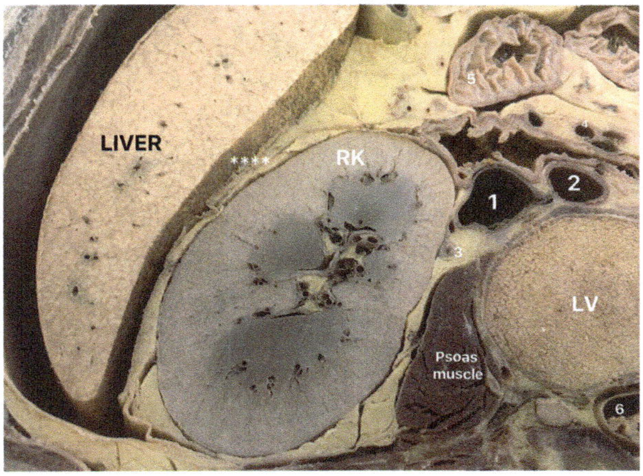

Fig. 3.2 The figure shows an inferior view of a transverse section at the level of the first lumbar vertebra (LV). We can observe the relationships between the right kidney (RK) and the psoas major muscle. In this figure, we also observe the relationships between the right kidney and the liver and the hepatorenal recess—the Morrison's Pouch (****). 1, inferior vena cava; 2, aorta artery; 3, ureter; 4, superior mesenteric vessels; 5, duodenum; 6, spinal medulla

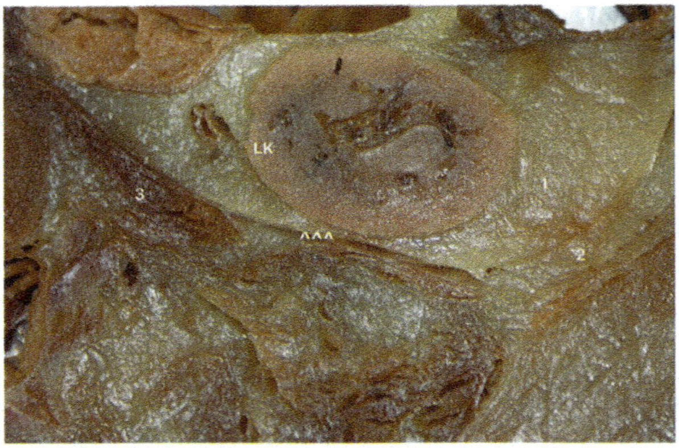

Fig. 3.3 Perirenal fat. The figure shows an inferior view of a transverse section at the level of the transition between the first and second lumbar vertebra showing the left kidney (LK) and the peri-renal fat (1). We can observe the relationship between the renal fascia (arrowheads) and the peri and para-renal fat (2); 3, psoas major muscle

fat. The anterior and posterior layer of renal fascia (Gerota's fascia) subdivides the retroperitoneal space into three potential compartments (Fig. 3.4): (a) the posterior para-renal space, which contains only fat; (b) the peri-renal space, which contains the adrenal glands, the kidneys, and proximal ureters, along with the peri-renal fat, and (c) the anterior para-renal space, which, in contrast with the intermediate and posterior spaces, is located in both sides of the median line in the abdomen and contains the ascending and descending colons, the second portion of the duodenum, and the pancreas (Sampaio 1996).

The superficial portion of the peri-renal fascia is continuous with the lateroconal fascia, which will itself fuse (black arrow) with the parietal peritoneum in front and form the lateral border between the anterior pararenal space and posterior pararenal space (Coffin et al. 2015). The lateroconal fascia delineates the lateral portion of the retroperitoneum (Fig. 3.4).

Superiorly, both renal fascia layers fuse above the adrenal gland and end united to the infra-diaphragmatic fascia. An additional layer of fascia separates the adrenal gland and the kidney (Sampaio 1996). In an interesting study, Qi and colleagues (Qi et al. 2015) show that the right anterior renal fascia and the right posterior renal fascia did not fuse above the kidney, indicating that the right renal fascia extended up into the bare area of the liver. In the left, the anterior renal fascia in some cases fuses with the posterior parietal peritoneum upwardly, including the splenorenal ligament and the omental sac, but in about 40% of the cases, the authors show that the left anterior renal fascia fused with the splenorenal ligament and the medial part fused with the inferior phrenic fascia, rather than fusion with the peritoneum of the omental sac (Fig. 3.5).

Laterally, both layers of the renal fascia fuse behind the descending and ascending colon. Medially, the posterior fascial layer fuses to that of the spinal muscles.

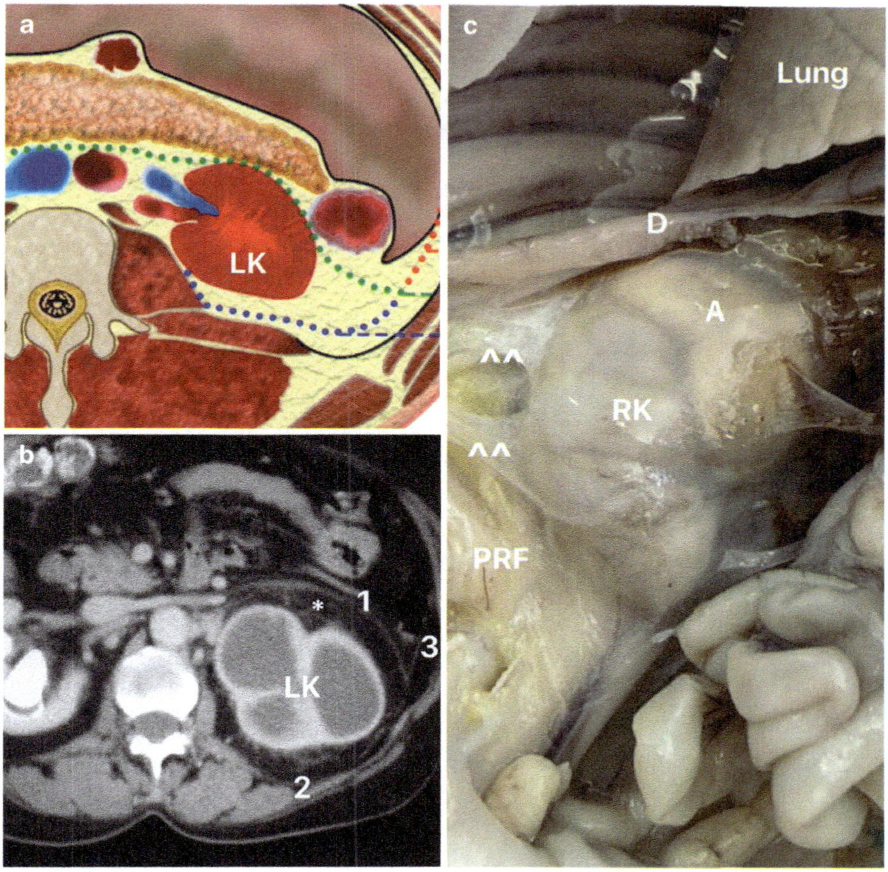

Fig. 3.4 Renal fascia and retroperitoneum spaces. (**a**) Schematic drawing showing the two parts of renal fascia: the anterior lamina (Gerota's fascia, green dotted line) and the posterior lamina (Zuckerkandl's fascia, blue dotted line). The superficial portion of the pararenal fascia is continuous with the lateroconal fascia (red dotted line). The lateroconal fascia delineates the lateral portion of the retroperitoneum. (**b**) Abdominal computerized tomography of a patient with hydronephrosis and infection; in this case, the anterior lamina of renal fascia (1), the posterior lamina of renal fascia (2), the lateroconal fascia (3), and the perirenal space are quite evident in this image examination. (**c**) The figure shows the abdominal cavity of a human fetus with 17 weeks post-conception; we can observe that the peritoneal organs are removed and the retroperitoneum is highlighted. The figure shows the renal fascia (arrowhead), the perirenal fat (PRF) involving the right kidney (RK), and the adrenal gland (A)

The anterior fascial layer mixes with the connective tissue of the great vessels (aorta and inferior vena cava). Inferiorly, the layers of renal fascial weakly fuse around the ureter, presenting a cone shape (Fig. 3.6). The renal fascia is not closed in its lower part, presenting communication with the posterior muscles of the abdomen, which explains in certain cases disseminations of collections in the perirenal space to the

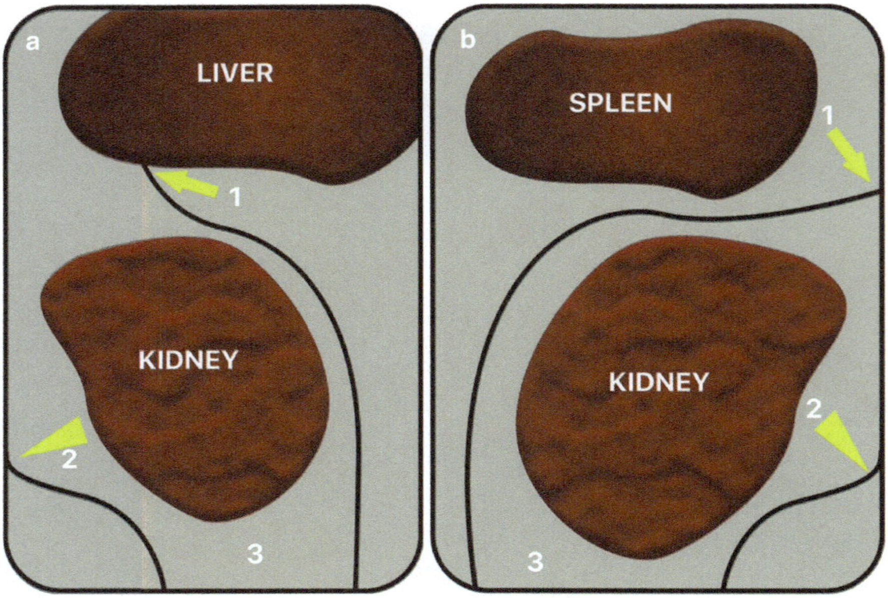

Fig. 3.5 Superior aspect of renal fascia. (**a**) Right renal fascia: schematic drawing showing the anterior renal fascia fused with the hepatorenal ligament (1) and the posterior renal fascia (2) fused with inferior phrenic fascia; in this situation, the right perirenal space communicated with the bare area of the liver; 3, perirenal space. (**b**) Left renal fascia: schematic drawing showing the anterior renal fascia fused superiorly with the inferior phrenic fascia (1) and the posterior renal fascia fused with the psoas fascia (2); in this situation, the perirenal space could not connect with the subdiaphragmatic retroperitoneal space; 3, perirenal space

psoas muscle, but the conic aspect of lower renal fascia explains why the perirenal disease rarely extends into the pelvic cavity and an open cone has not been observed on CT scans that were studied in the previous study of the anatomy of the caudal extent of the cone of the renal fasciae in cadavers and on CT scans (Raptopopoulous et al. 1995) (Fig. 3.7).

The kidneys rest over the major psoas and square lumbar muscle (Fig. 3.8). Usually, the right kidney is higher than the left one, and the posterior surface of the right kidney is crossed by the twelfth rib, and the posterior surface of the left kidney is crossed by the eleventh and twelfth ribs. The posterior surface of the diaphragm is united to the ends of the eleventh and twelfth ribs. Close to the vertebral spine, the diaphragm is connected to the muscles of the posterior wall of the abdomen, forming the arched medial and lateral ligaments on each side. Therefore, the posterior surface of the diaphragm is arched as a dome over the superior pole of both kidneys (Netter 1978; Sampaio 1996).

The relationship between the kidney and the diaphragm is of great importance in performing and especially in accessing percutaneous renal surgeries (Hooper et al.

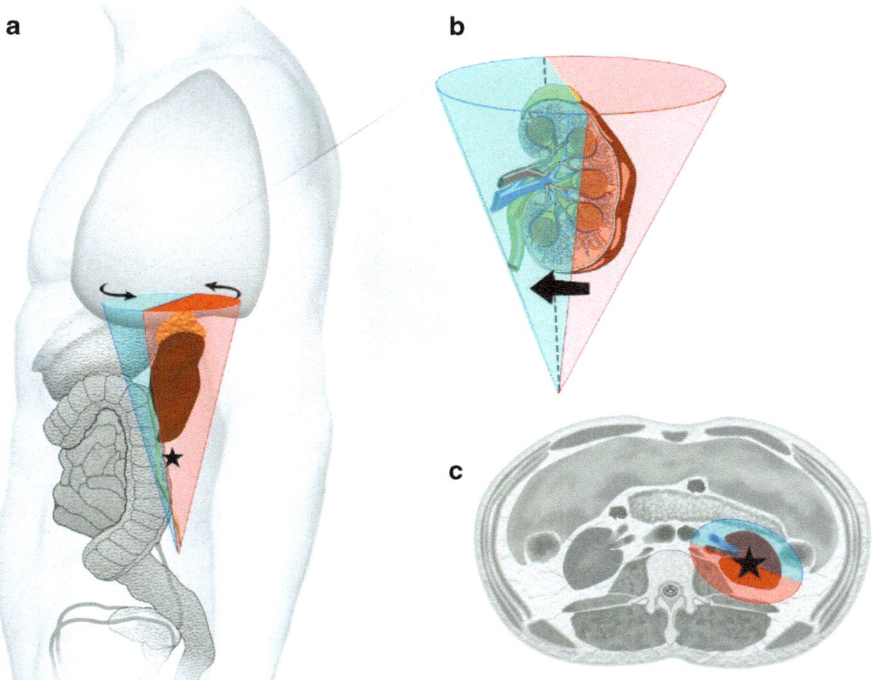

Fig. 3.6 Inferior aspect of renal fascia. (**a**) Schematic drawing of the abdomen showing the disposition of the inferior portion of renal fascia. The posterior layer of the renal fascia is in red, and the anterior layer of the renal fascia is in blue. The inferior portion of the renal fascia has a cone shape (*). (**b**) Schematic drawing showing only the relationship of renal fascia with the kidney; the arrow shows the inferior aspect of renal fascia. (**c**) Schematic drawing of a sectional view at the level of the first lumbar vertebra showing the left kidney (*) and the two layers of the renal fascia—posterior layer in red and anterior layer in blue

1987; Chow et al. 2020). In the elegant study of Chow (Chow et al. 2020), the author concluded that the decreased distance between the posterior insertion of the diaphragm (medial and lateral arcuate ligaments) and the superior edge of the renal upper pole on preoperative CT scan was associated with TCs from supracostal puncture during PNL. Critical preoperative recognition of this anatomic relationship can help preoperative planning and patient counseling and may prevent or reduce TCs. In Fig. 3.9, we can observe the relationships between the superior renal pole and the diaphragm.

The kidney maintains relations with the liver, spleen, and colons (Figs. 3.10 and 3.11). Kidney relation with the liver is very important in urology and radiology. The space between the right kidney and liver is called hepatorenal recess (Morison space or Morison's pouch), which is well visualized in image exams, particularly in the presence of free liquid in the abdominal cavity (Fig. 3.10). Due to their dimensions,

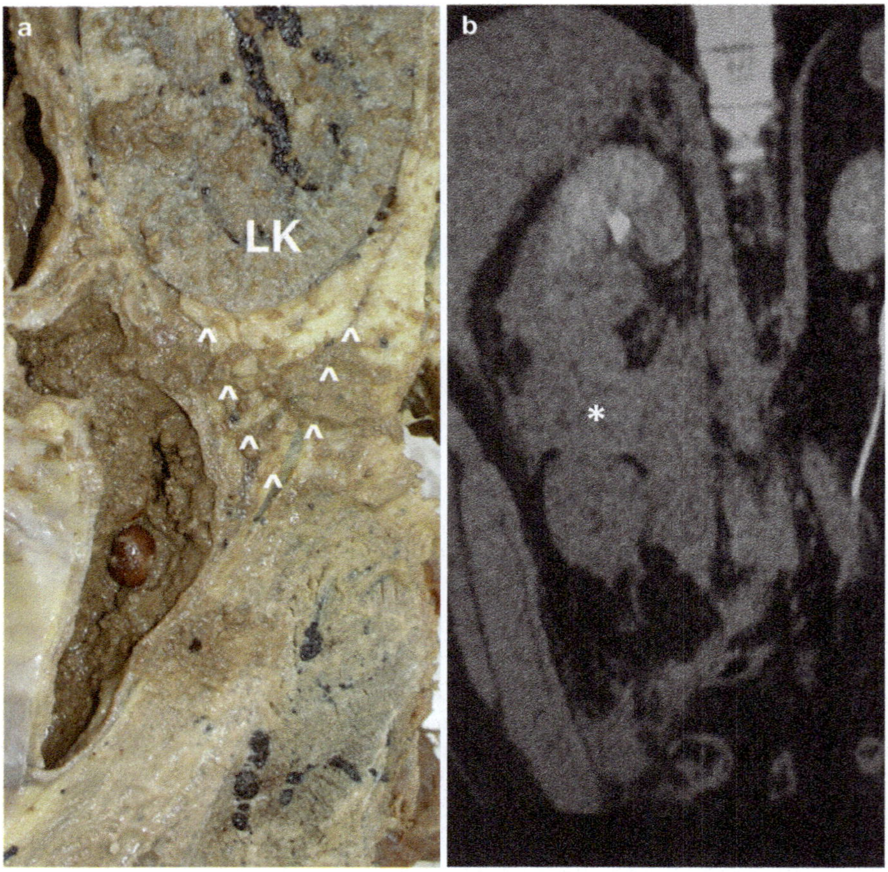

Fig. 3.7 Inferior aspect of renal fascia. (**a**) The figure shows a frontal section of the lower part of the abdomen; we can observe the left kidney (LK) and the inferior portion of renal fascia with this aspect of a cone shape (arrowhead). (**b**) Abdominal computerized tomography of a patient with renal trauma showing the dissemination of the hematoma (*) inferiorly

the liver and spleen may be positioned posterior and laterally at the level of the supra-hilar region of the kidney. Therefore, in endourology, it is important to observe that the space for a high renal puncture is very limited. If the puncture is performed while the patient is medium or intensively breathing, there is a high risk of lesion of the liver or spleen. This knowledge is particularly important in patients with hepatomegaly or splenomegaly (Sampaio 1996). In those cases, it is important to obtain previously a computer tomography before puncturing the kidney.

The descending colon extends from the ileum-colic valve until the right colonic flexure (hepatic flexure), becoming the transverse colon. The hepatic colonic flexure

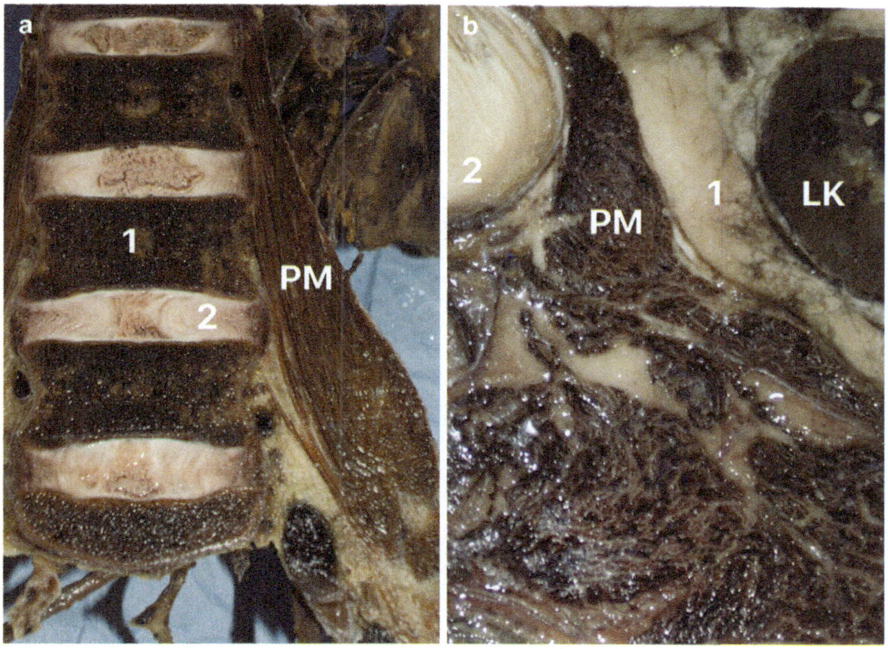

Fig. 3.8 Psoas major muscle. (**a**) The figure shows a frontal section of a formalized corpus showing the position of the psoas major muscle (PM); 1, third vertebral body; 2, intervertebral disc. (**b**) The figure shows an inferior view of a transverse section of the retroperitoneum at the level of the second lumbar vertebra; we can observe the relationship between the left kidney (LK) and the psoas major muscle (PM); 1, perirenal fat; 2, intervertebral disc

(hepatic angle) leans anteriorly to the inferior portion of the right kidney (Fig. 3.11). The descending colon extends inferiorly to the left colonic flexure (splenic flexure) at the level of the iliac crest. The left colonic flexure leans anteriorly and laterally to the left kidney. It is important to consider the retroperitoneal position of the ascending and descending colons. During routine computer tomography (CT) exams, it was observed the retroperitoneal colon leaning in a posterior and lateral or retro-renal position. In those cases, there is an increased risk of the lesion during percutaneous intra-renal access. The retro-renal colon is more frequently related to the lower kidney poles (Williams et al. 1995). In a study controlled by CT, the retro-renal colon was observed in 1.9% of cases, while the patients were in a supine position. However, when in a ventral position (more frequently position used for percutaneous renal access), the retro-renal colon was observed in 10% of patients (Hooper et al. 1987). Therefore, in all renal puncture procedures, it is important to carefully analyze the fluoroscopic exam with the patient in ventral decubitus in order to detect the presence of a retro-renal colon. The fluoroscopic exams are particularly important in the inferior polar region of the kidneys.

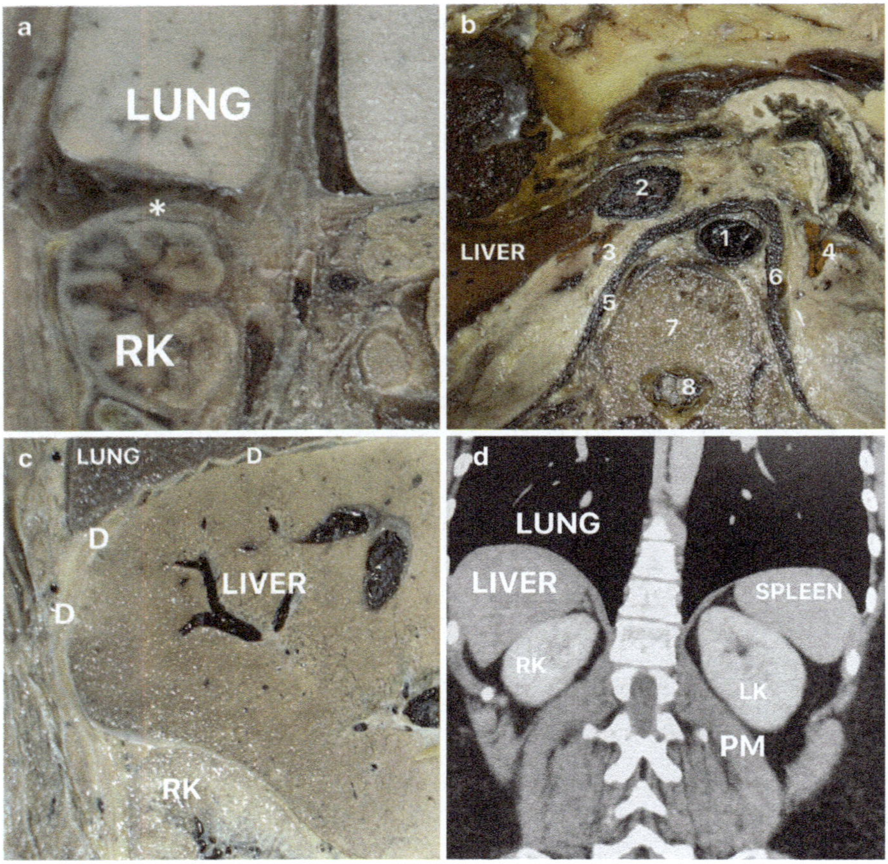

Fig. 3.9 Relationships between the superior renal pole and the diaphragm. (**a**) The figure shows a frontal section of a human fetus of the second gestational trimester; we can observe the relationship between the upper pole of the right kidney (RK) with the diaphragm (*). (**b**) Transverse section at the level of the eleventh thoracic vertebra showing the relationships of the diaphragm with the retroperitoneum space; we can observe the important relationship between the aorta artery (1) with the right (5) and the left (6) diaphragm pillars; 2, inferior vena cava; 3, right adrenal gland; 4, left adrenal gland; 7, vertebral body; 8, spinal medulla. (**c**) Frontal section of the superior abdomen showing the relationships between the right kidney (RK) upper pole with the diaphragm (D), liver, and right lung. (**d**) In this figure, we can observe a computerized tomography of the superior abdomen showing the relationships of the upper pole of the right (RK) and left (LK) kidneys with the diaphragm; PM, psoas muscle

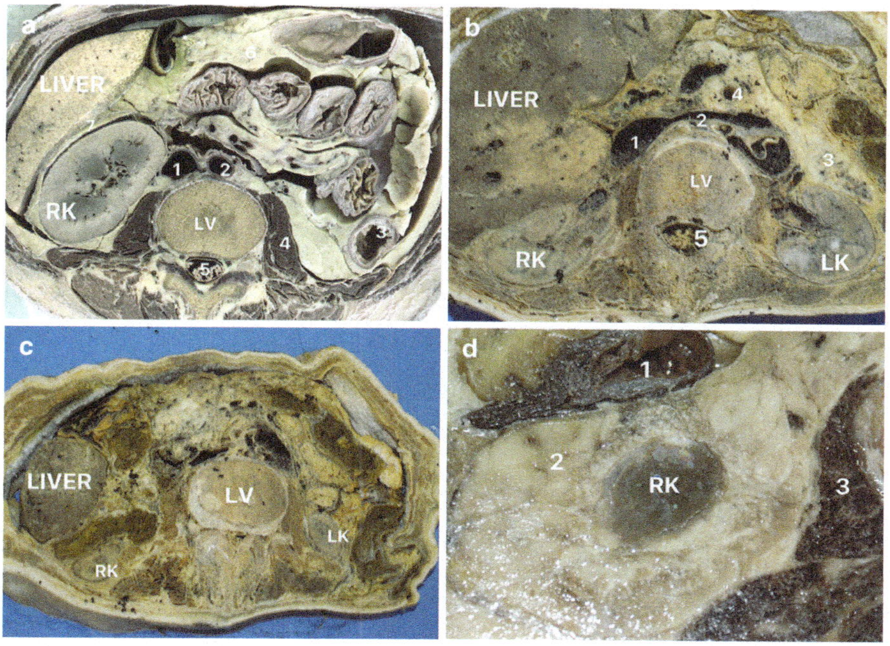

Fig. 3.10 Right kidney relationships. (**a**) Inferior view of a transverse section of the abdomen at the level of the transition between the twelfth thoracic vertebra and the first lumbar vertebra; we can observe the right kidney (RK), superior vena cava (1), aorta artery (2), left colon (3), and psoas major muscle. (**b**) Inferior view of a transverse section of the abdomen at the level of the first lumbar vertebra; we can observe the right kidney (RK), the left kidney (LK), superior vena cava (2), aorta artery (2), pancreas (3), and superior mesenteric artery. (**c**) Inferior view of a transverse section of the abdomen at the level of the transition between the first and the second lumbar vertebra; we can observe the right kidney (RK) and the left kidney (LK). (**d**) Inferior view of a transverse section of the abdomen at the level of the transition of the first and second lumbar vertebra; we can observe the right kidney (RK), the right colon (1), the perirenal fat (2), and the psoas major muscle (3)

3.1.2 Vascular and Lymphatic Anatomy of the Kidney

The renal artery is a branch of the lateral aspect of the abdominal aorta, which origins around 1.5 cm below the superior mesenteric artery. The right renal artery reaches the renal hilum and runs posteriorly to the inferior vena cava. The left renal artery is shorter than the right one. There are many frequent variations of the main renal artery. It is more common to observe multiple renal arteries than renal veins. The main renal artery is also more prevalent than any other arteries with the same caliber. In a great series of renal pedicles analyzed, it was observed variations of the renal artery in 30% of patients (Weld et al. 2005; Sampaio and Favorito 1993; Sampaio 1993).

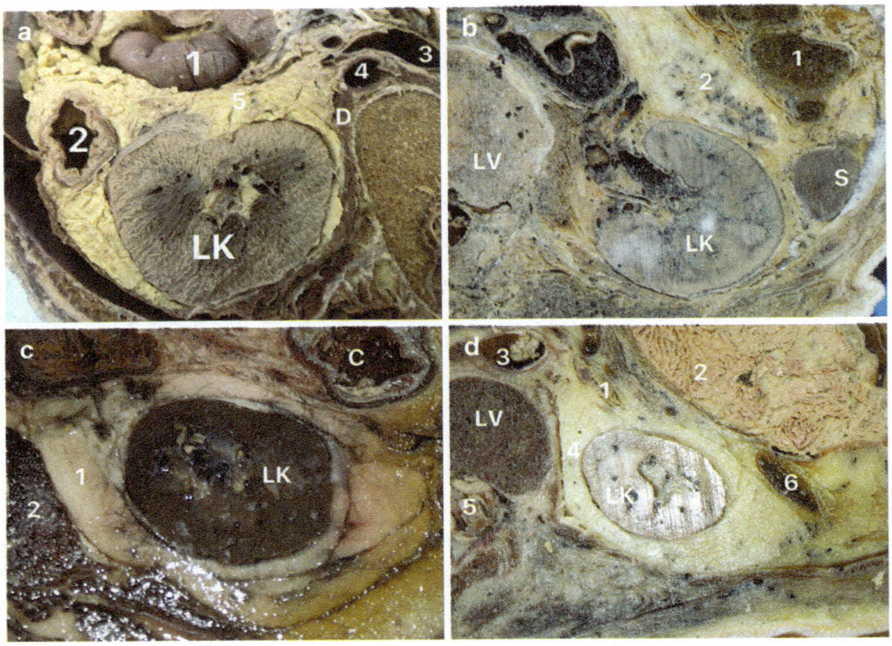

Fig. 3.11 Left kidney relationships. (**a**) Superior view of a transverse section of the abdomen at the level of the first lumbar vertebra; we can observe the left kidney (LK), the diaphragm (D), transverse colon (1), left colon (2), superior vena cava (3), and aorta artery (4). (**b**) Inferior view of a transverse section of the abdomen at the level of the first lumbar vertebra (LV); we can observe the left kidney (LK), spleen (S), the left colon (1), and the pancreas (2). (**c**) Inferior view of a transverse section of the abdomen at the level of the transition between the first and the second lumbar vertebra; we can observe the left kidney (LK), the perirenal fat (1), psoas major muscle (2), and left colon (C). (**d**) Inferior view of a transverse section of the abdomen at the level of the first lumbar vertebra (LV); we can observe the left kidney (LK), adrenal gland (1), the stomach (2), aorta artery (3), the perirenal fat (4), the spinal medulla (5), and the splenic flexure of the colon (6)

The main renal artery divides into an anterior and a posterior branch, after originating the inferior adrenal artery. The posterior branch (retro-pyelic artery) continues as a posterior segment artery that irrigates the posterior homonym segment without significant ramifications, and the anterior branch of the renal artery divides into three or four segment arteries (Fig. 3.12). Before entering the renal parenchyma, the segment arteries divide into interlobar arteries (infundibular arteries) that run adjacent to the calix infundibulum and the smaller calix, reaching the renal columns between the renal pyramids (Sampaio 1996).

As the interlobar arteries progress, close to the base of the pyramids, they originate the arched arteries (usually by dichotomy). The arched arteries originate the interlobular arteries that run peripherally, forming the glomerular afferent arterioles (Sampaio 1996; Sampaio and Favorito 1993).

On the other side, the intra-renal veins do not present a segment pattern. Also, contrary to the arteries, there are free communications along the venous system,

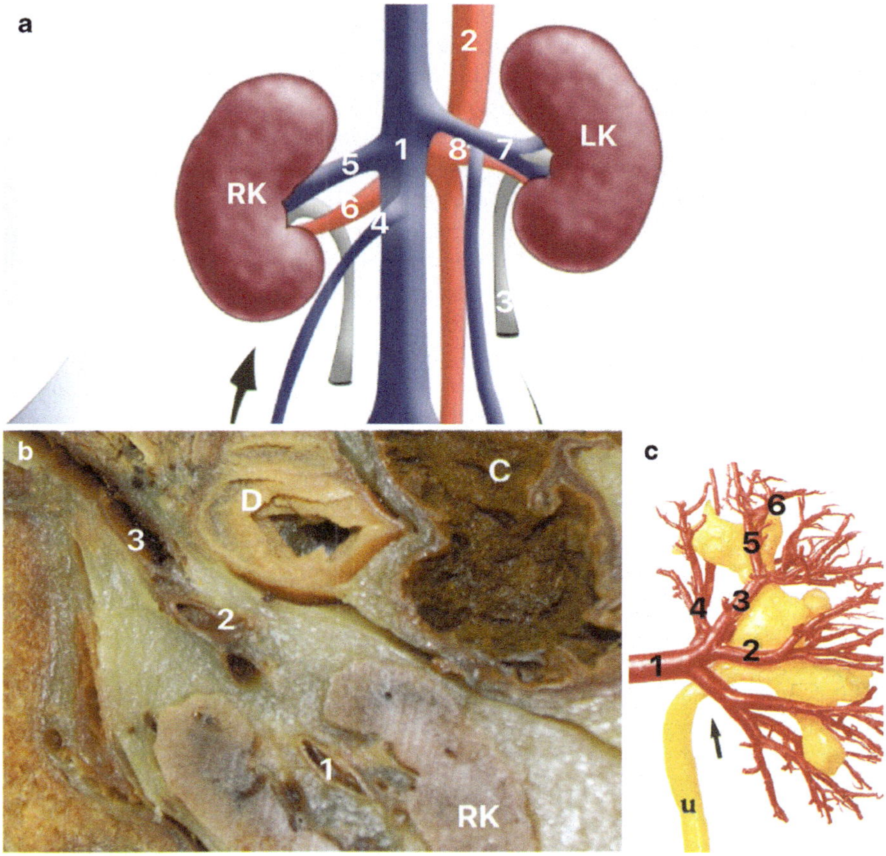

Fig. 3.12 Vascular anatomy of the kidney. (**a**) Schematic drawing showing the arteries and veins of the kidney: (1) inferior vena cava, (2) aorta artery, (3) ureter, (4) right gonadal vein, (5) right renal vein, (6) right renal artery, (7) left renal vein, and (8) left renal artery; the arrow shows the flow direction in the gonadal vein. (**b**) Superior view of a transverse section of the abdomen at the level of the first lumbar vertebra (LV); we can observe the right kidney (RK), duodenum (D), right colon (C), renal pelvis (1), right renal artery (2), and right renal vein (3). (**c**) The figure shows an endocast of the kidney arterial system and collector system; we can observe renal artery (1), the segmental arteries—the arrow showing the inferior segment artery, the anteroinferior segment artery (2), the anterosuperior segment artery (2), and the superior segment artery (3); the segment arteries divide into interlobar arteries (infundibular arteries) (5) and the interlobar arteries (6)

with broad anastomosis among veins. Therefore, these anastomosis prevent congesting and ischemia of the renal parenchyma in the presence of venous lesions. The right renal vein is shorter and usually without branches, while the left renal vein is longer, running posteriorly to the aorta, usually receiving blood from three tributaries: gonadal, adrenal, and lumbar veins (Sampaio 1993).

Intra-renal lymphatics are divided into superficial and deep plexus. The superficial plexus is located right after the renal capsule and is attached to cortical lymphatic vessels. In pathologic conditions (for example, pyelonephritis), this plexus may communicate with an extra-renal plexus, located at the peri-renal fat, which drains to aortic lymph nodes in the lumbar region (Tanagho and Smith 1968). The deep plexus is subcortical and also pyramid-shaped (located more profoundly). These deep plexus are peri-vascular and drain together, running along with the arched and interlobar vessels, converging to the renal hilum. The collecting channels arise at the renal hilum, and if the renal artery is present, usually the lymphatic vessel will follow that one. There are one to four lymphatic channels that arise on the renal hilum, anterior or posteriorly to the renal vein. These lymphatic channels may anastomose and show a plexiform aspect. Usually, they are peri-arterial, forming an anterior plexus, when arising at the ventral surface of the kidney, or a posterior plexus, when arising at the dorsal surface of the kidney. In some cases, the lymphatic channels may directly connect to their lymph nodes without following the arterial branches (Tanagho and Smith 1968).

The lymphatic vessels of the left kidney may divide into posterior and anterior branches. The posterior channels leave the renal hilum and run dorsally to the renal vessels to reach the lymph nodes of the pillar of the diaphragm muscle. The anterior channels run ventrally to the renal vein, reach the lymph nodes located above or below the origin of the renal artery, draining the superior or lower pole of the kidney, respectively. There are also lymph nodes that origin at the inferior pole of the kidney to reach the lymph nodes located at the origin of the spermatic artery, at the lateral surface of the aorta (Assouad et al. 2006).

It is important to stress that in both sides, the lymphatic drainage of the posterior region of the kidney (dorsal surface) reaches directly the lymph nodes of the diaphragmatic pillars, in contact with the hiatus of the splanchnic nerve. From that point on, through the diaphragm, these lymphatic vessels drain to the retro-aortic lymph nodes from T11 to L1 (mediastinal lymph nodes). This detail is fundamental when analyzing the dissemination of metastasis of renal cell carcinoma (Assouad et al. 2006).

3.1.3 Adrenal Glands

The adrenal (supra-renal) glands are endocrine, yellowish, and small glands, weighing 3 to 5 grams. During the fetal period, the adrenal glands are very big, with the size of 1/3 of the kidney (Fig. 3.13). After birth, the cortical region degenerates and the glands shrink. It has two regions: the peripheric cortex, responsible for the production of glucocorticoids and mineralocorticoids, and the central medullar, responsible for the production of adrenaline and noradrenaline (Netter 1978).

They are located between the superior and medial face of the anterior surface of the kidney and the diaphragm. The glands are surrounded by the renal fascia, are located in the peri-renal space, and are involved by fat. The right gland is pyramid-shaped, and the base is in contact with the kidney, and the posterior surface is in contact with the diaphragm muscle and inferior vena cava. The left adrenal is

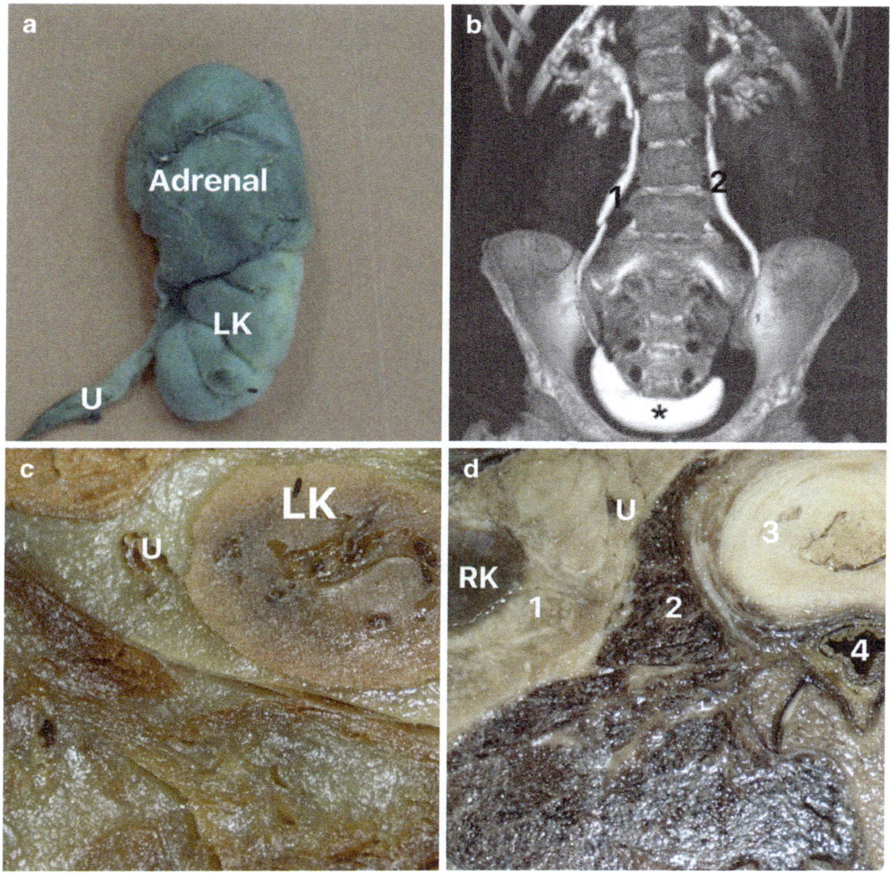

Fig. 3.13 Adrenal gland and ureter. (**a**) Fetus with 20 weeks post-conception (WPC); we can observe that the adrenal glands are very big, measuring 1/3 of the kidney; RK, right kidney; U, ureter. (**b**) Abdominal computerized tomography of abdomen showing the ureters; 1, right ureter; 2, left ureter; *, bladder. (**c**) Inferior view of a transverse section of the abdomen at the level of the transition between the twelfth thoracic vertebra and the first lumbar vertebra; we can observe the left ureter (U) inside the perirenal fat; LK, left kidney. (**d**) Inferior view of a transverse section of the abdomen at the level of the first lumbar vertebra; we can observe the right ureter (U), inside the perirenal fat (1); RK, right kidney; 2, psoas major muscle; 3, lumbar vertebra; 4, spinal medulla

semi-lunar shaped and is posteriorly related to the diaphragm and anteriorly to the pancreas, spleen, and stomach (Williams et al. 1995).

The adrenal is irrigated by three arteries and drained by only one vein. The knowledge of the vascular anatomy of the adrenals is of great practical importance during their surgical removal, particularly by laparoscopy. The three arteries that irrigate the gland are superior adrenal (branch of the lower phrenic); medial adrenal, a direct branch of the aorta; and the inferior adrenal, a branch of the renal artery. The arterial distribution is the same on both sides (Williams et al. 1995).

There is only one vein that drains the adrenal, the adrenal vein, which is, at right, an affluent of the inferior vena cava and, which, at left, drains to the left renal vein. The adrenal vein shows a relatively large caliber and is one of the main structures that must be ligated during resection surgeries of the gland (Netter 1978; Williams et al. 1995).

3.1.4 Ureter

The ureter is a tubular conduit made of smooth muscle fibers with peristalsis, which connects the renal pelvis to the bladder. It is 22–30 cm long (Fig. 3.13) and internally presents a mucous layer composed of pseudo-stratified epithelium, similar to the bladder (Wolf et al. 1996; Abdel-Razzak 1996). The muscular layer presents three layers: circular fibers (more internally), internal longitudinal fibers, and external longitudinal fibers. Externally, it is covered by an adventitia layer continuous to the renal capsule.

The entire ureter is located retroperitoneally and may be divided into two portions: abdominal and pelvic. Its abdominal portion is closely related to the major psoas muscle and its sheath; it runs anteriorly and medially in relation to that muscle and to the genital-femoral nerve, with the peritoneum located ahead (Fig. 3.13). Psoas muscle hypertrophy may cause alteration of the ureteral path, with a lateral detour of the ureter below the renal pelvis and a medial detour under the fifth lumbar vertebra (Bree et al. 1976).

At the right, the ureter is close to the inferior vena cava and relates to the right colic vessels and terminal ileum before entering the pelvis. The left ureter is covered by the initial portion of the jejunum in the beginning of its path, and it is related to the aorta and left colic vessels. When approaching the pelvis, it crosses the sigmoid and its mesocolon. Ureteral relations with the colonic vessels, and particularly the sigmoid, are very important to be observed during intestinal resection surgeries. In its abdominal path, on both sides, the ureter is lateral to gonadal vessels (Kabalin 1992; Redman 1996). Before entering the pelvis, the ureter crosses anteriorly the iliac vessels. Therefore, endourological manipulation of calculi in that region by ureterorenoscopy should be made carefully since ureteral perforation in that region may lead to severe vascular lesions.

The ureter is irrigated by several arteries along its path. The main arteries responsible for the irrigation of the abdominal portion are the renal, gonadal, aorta, and common iliac arteries (Williams et al. 1995). The branches of the renal artery that irrigate the ureter arise close to the renal sinus. The arteries responsible for the irrigation of the pelvic portion of the ureter include the internal iliac and some of its branches, such as the vesical inferior, uterine, medium rectal, and vaginal (Netter 1978). Anomalous vessels may cause obstruction of the pelvic ureter. Variations of the normal path of some branches of the internal iliac artery (mainly inferior vesical and umbilical) may cause obstruction of the ureterovesical junction and hydronephrosis.

Some aspects with great surgical interest are relevant, such as the fact that the arteries that irrigate the abdominal ureter reach it at its medial wall, while those that irrigate the pelvic ureter reach it at its lateral wall. All arterial branches, after reaching the ureter, have a longitudinal path in the interior of the ureteral adventitia, which form a major anastomotic plexus. Therefore, the surgical dissection of the ureter should preserve the adventitia as much as possible to avoid secondary lesions due to ischemia (Kabalin 1992).

The lymphatic and veins go along the arteries. In the pelvis, the lymphatic vessels drain preferably to the internal, external, and common iliac lymph nodes. In the abdomen, the left ureteral lymphatic vessels drain to the left para-aortic lymph nodes, while the right ureteral lymphatic vessels drain to the inter-aorto-cava and right para-cava lymph nodes. The lymphatic vessels of the superior portion of the ureter and renal pelvis usually drain the lymphatic vessels of the homolateral kidney. Therefore, the lymphatic dissemination of ureteral lesions depends on the stricken lesion (Redman 1996). The ureter presents an abundant autonomous innervation. Ureteral nerves originate at the celiac, aortic-renal, mesenteric, and superior and inferior hypogastric plexuses (Kabalin 1992). The exact function of the autonomous innervation of the ureter is still not clearly known.

References

Abdel-Razzak OM. Ureteral anatomy. In: Smith's textbook of endourology. Quality Medical Publishing: New York - USA; 1996. p. 388.

Assouad J, Riquet M, Foucalult C, Hidden G, Delmas V. Renal lymphatic drainage and thoracic duct connections: implications for cancer spread. Lymphology. 2006;39:26–32.

Bree RL, Grenn B, Keiller DL, Genet EF. Medial deviation of the ureter secondary to psoas muscle hypertrophy. Radiology. 1976;118:691–5.

Chow AK, Ogawa S, Seigel C, Sands KG, Vetter J, Desai A, Venkatesh R. Evaluation of perirenal anatomic landmarks on computed tomography to reduce the risk of thoracic complications during Supracostal percutaneous nephrolithotomy. J Endourol. 2020; https://doi.org/10.1089/end.2020.0551, online ahead of print.

Coffin A, Boulay-Coletta I, Sebbag-Sfez D, Zins M. Radioanatomy of the retroperitoneal space. Diagn Interv Imaging. 2015;96:171–86.

Hooper KD, Sherman JL, Luethke JM. The retrorrenal colon in the supine and prone patient. Radiology. 1987;162:443–6.

Kabalin JN. Surgical anatomy of genitourinary tract - anatomy of the retroperitoneum and kidney. In: Campbell's urology. 6th ed. New York: Saunders; 1992.

Mindell HJ, Mastromatteo F, Dickey KW, et al. Anatomic communications between the three retroperitoneal spaces: determination by CT-guided injections of contrast material in cadavers. AJR. 1995;164:1173–8.

Molmenti EP, Balfe DM, Kanterman RY, Bennett HF. Anatomy of the retroperitoneum: observations of the distribution of pathologic fluid collections. Radiology. 1996;200:95–103.

Netter F. Reproductive system, vol. 2. New Jersey: Ciba; 1978.

Ochi A, Satoru M, Adachi T, Akita K. Zoning inside the renal fascia: the anatomical relationship between the urinary system and perirenal fat. Int J Urol. 2020;27:625–33.

Qi R, Yu JQ, Zhou XP, Li ZL. The superior aspect of the perirenal space: could it be depicted by dual-source CT in vivo in adults. Br J Radiol. 2015;88(1045):20140480.

Raptopopoulous V, Lei QF, Touliopoulos P, et al. Why perirenal disease does not extend into the pelvis: the importance of closure of the cone of the renal fasciae. AJR. 1995;164:1179–84.

Redman JF. Anatomy of the genitourinary system. In: Adult and pediatric urology. 3rd ed. St. Louis: Mosby - Year Book; 1996.

Sampaio FJB. Renal arterial pedicle: anatomic analysis applied to urologic and radiologic procedures. In: Sampaio FJB, Uflacker R, editors. Renal anatomy applied to urology, endourology and interventional radiology. New York: Thieme Medical; 1993. p. 47.

Sampaio FJB. Anatomy of the kidney for endourology. In: Segura JW, editor. Smith's textbook of endourology. Part II: percutaneous surgery. St. Louis: Quality Medical; 1996. p. 150.

Sampaio FJB, Favorito LA. Ureteropelvic junction stenosis: vascular anatomical background for endopyelotomy. J Urol. 1993;150:1787–981.

Tanagho EA, Smith DR. Mechanism of urinary continence. I. Embryologic, anatomic and pathologic considerations. J Urol. 1968;100:640–6.

Weld KJ, Bhayani SB, Belani J, Ames CD, et al. Extrarenal vascular anatomy of kidney: assessment of variations and their relevance to partial nephrectomy. Urology. 2005;66(5):985–9.

Williams PL, Bannister LH, Berry MM, Dyson M, Dussek JE, Ferguson MWJ. Gray's anatomy. 38th ed. Churchill-Livingstone: Edinburgh; 1995.

Wolf JS, Humphrey PA, Rayala HJ, Gardner SM, Mackey RB, Clayman RV. Comparative ureteral anatomy. J Endourol. 1996;10:527–31.

Chapter 4
Sectional Anatomy of the Male Pelvis

Luciano Alves Favorito, Natasha T. Logsdon, and Francisco J. B. Sampaio

4.1 Introduction

The male genital organs are frequently afflicted by pathologies, and the knowledge of their anatomy is highly important to understand the pathologies and the interpretation of the image exams, especially in MRI due to the advancement and importance of the prostate imaging, shortly after implementation (PI-RADS) in prostate cancer diagnosis (Skilinda et al. 2019). In this chapter, we will show some aspects of the sectional anatomy of the bladder, prostate, penis, and scrotum applied to image exams and urological pathologies.

L. A. Favorito (✉)
Urogenital Research Unit, Rio de Janeiro State University, Rio de Janeiro, RJ, Brazil

IDOMED-UNESA, Rio de Janeiro, RJ, Brazil

The National Council for Scientific and Technological Development (CNPq), Lago Sul, Federal District, Brazil

The Rio de Janeiro State Research Foundation (FAPERJ), Rio de Janeiro, RJ, Brazil

N. T. Logsdon
University Center Geraldo di Biasi, Rio de Janeiro, RJ, Brazil

Urogenital Research Unit, Rio de Janeiro State University, Rio de Janeiro, RJ, Brazil

F. J. B. Sampaio
Urogenital Research Unit, Rio de Janeiro State University, Rio de Janeiro, RJ, Brazil

The National Council for Scientific and Technological Development (CNPq), Lago Sul, Federal District, Brazil

The Rio de Janeiro State Research Foundation (FAPERJ), Rio de Janeiro, RJ, Brazil
e-mail: sampaio@uerj.br

L. A. Favorito, N. T. Logsdon (eds.), *Atlas of Sectional Anatomy*,
https://doi.org/10.1007/978-3-030-91688-6_4

4.2 Prostate and Bladder

In male pelvis, the bladder is covered superiorly by the peritoneum, which continues as an anterior parietal lamina on its ventral surface. On the dorsal face, the peritoneum covering the bladder merges with the visceral peritoneum of the vesical rectal recess (Brooks 2002) (Fig. 4.1). The anteroinferior and lateral faces of the bladder are separated from the pelvic walls by a space filled with fatty tissue and by the periprostatic venous plexus, the Retzius space (Brooks et al. 1998). The anatomical relationships between the bladder and the male pelvic viscera are very well

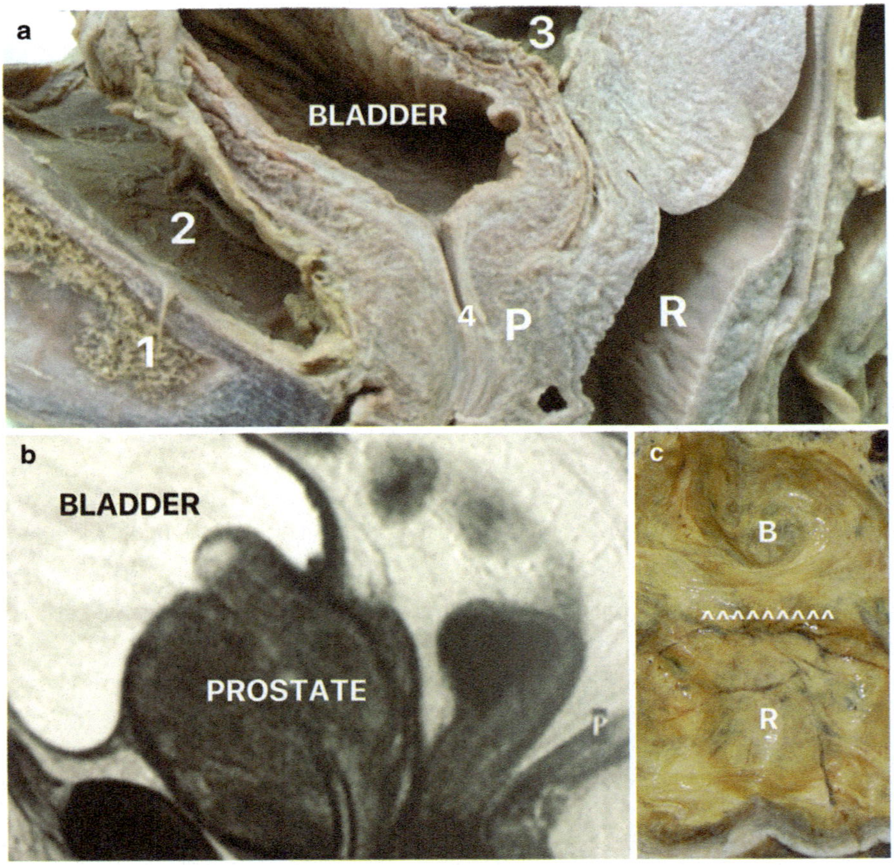

Fig. 4.1 (**a**) Sagittal section of a male pelvis from a fixed corpse indicating the relations of the bladder and the prostate (P); R, rectum; 1, pubis; 2, Retzius space; 3, vesical rectal reflection; 4, prostatic urethra. (**b**) Magnetic resonance image showing the relationships between bladder and prostate. (**c**) Transverse section of a male pelvis showing the relationships between the bladder (B) and rectum (R); the arrowhead represents the rectovesical pouch, and the peritoneum is lining the upper side of the pelvis

evaluated in imaging studies, especially by magnetic resonance imaging (MRI) (Fig. 4.1).

The bladder neck is continuous inferiorly to the base of the prostate. The prostate is frequently associated with benign diseases (benign prostatic hyperplasia and infections) and prostate cancer. The prostate has a base, apex, anterior face, posterior face, and two inferolateral faces. The urethra penetrates the central region of the prostate base. From below, the prostate apex rests on the upper surface of the urogenital diaphragm (Fig. 4.2). From behind, the posterior surface of the prostate is intimately related to the anterior surface of the rectum and separated from this by the rectoprostatic septum (Denonvilliers' fascia) (Brooks 2002).

From the front, the anterior surface of the prostate is related to the pubic symphysis, separated from it by the extraperitoneal fat in the retropubic space, which is covered by the pelvic fascia (Fig. 4.3). The endopelvic fascia is the outer layer of the parietal pelvic fascia (Raychaudhuri and Cahill 2008; Julio Junior et al. 2015). It covers the obturator internus, piriformis, levator ani, and coccygeus

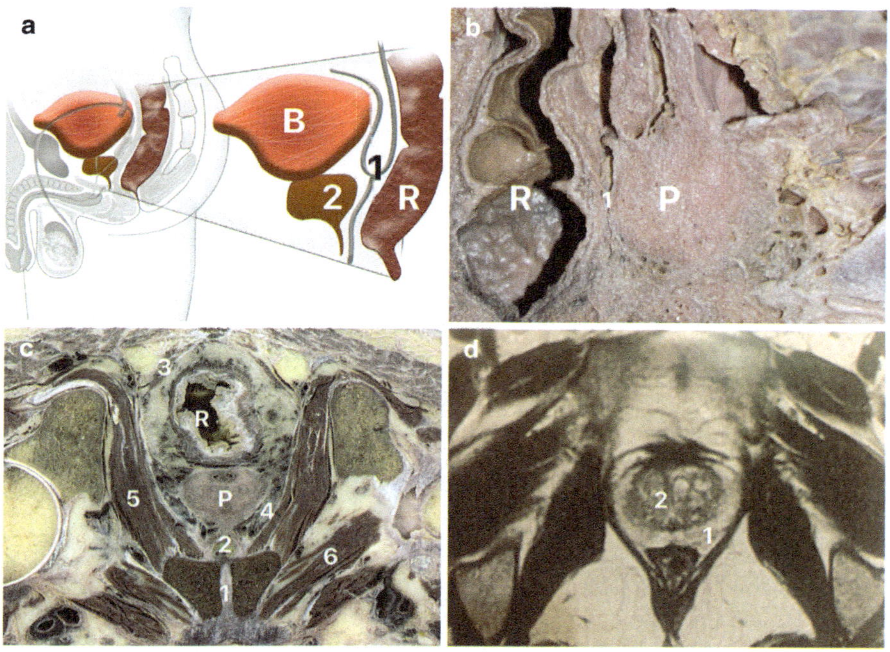

Fig. 4.2 (**a**) Schematic drawing showing the bladder (B) and rectum (R); 1, rectovesical pouch; 2, prostate. (**b**) Sagittal section of a male pelvis from a fixed corpse indicating the relationships between the prostate (P) and the rectum (R); 1, Denonvilliers fascia. (**c**) Transverse section of a male pelvis at the level of the pubic symphysis (1) showing the relationships of the prostate (P); R, rectum; 2, Retzius space; 3, levator anus muscle; 4, periprostatic venous plexus; 5, internal obturator muscle. (**d**) Magnetic resonance image of a male pelvis showing the prostatic zones; 1, peripheric zone; 2, central zone

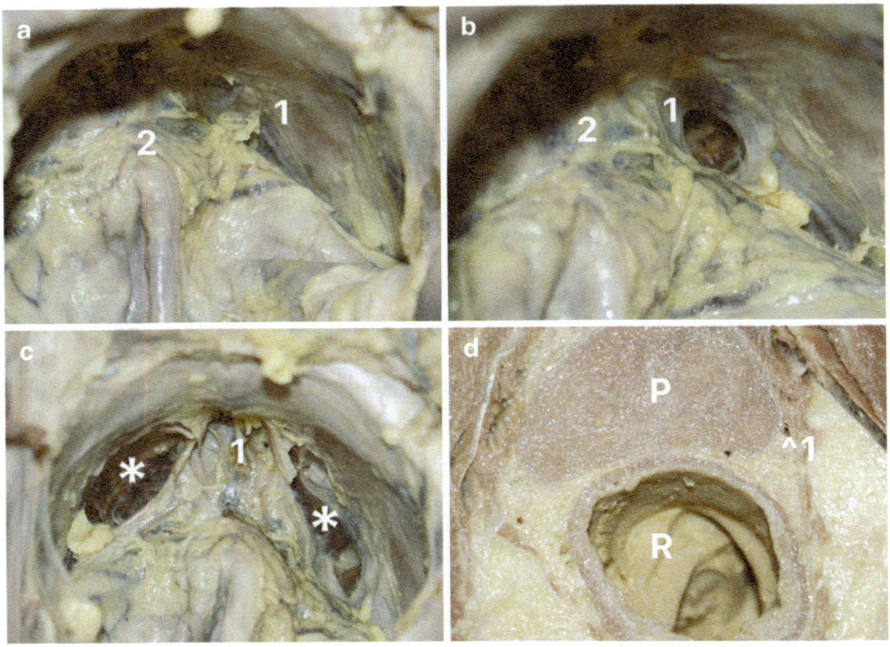

Fig. 4.3 Pelvic fascia in male. (**a**) Superior view of a male pelvis showing the endopelvic fascia (1); we can observe the periprostatic venous plexus (2). (**b**) In the same pelvis, the endopelvic fascia is opened on the right side (1); 2, periprostatic venous plexus. (**c**) The endopelvic fascia was opened bilaterally (*) and the venous plexus is isolated (1). (**d**) Transverse section of a male pelvis showing the relationships between the prostate (P) and the neurovascular bundle (1); R, rectum

muscles, and it is continuous with the transversalis fascia and fused to the periosteum of the pubic bone (Raychaudhuri and Cahill 2008). The prostatic fascia fuses with the endopelvic fascia. An important structure in the continence mechanism is the pubovesical/puboprostatic ligaments (Takenaka et al. 2005, 2007; Vora et al. 2013). They are paired fibrous structures extending from the outer longitudinal bladder muscle to the distal end of the posterior part of the pubic symphysis adjacent to the anterior margin of the internal urethral sphincter (Takenaka et al. 2007) (Fig. 4.3).

The endopelvic fascia is thin, overlying the levator ani, and the lateral pelvic fascia has an important relationship with the neurovascular bundle (NVB) (Hinata et al. 2013). The NVB is one of the most important structures to be preserved during radical prostatectomy (Walz et al. 2010). The NVB is situated laterally and in the posterolateral angle of the prostate and seminal vesicles as a bundle or spread out within adipose-connective tissue between layers of prostatic and levator ani fascia (Patel et al. 2012) (Fig. 4.3).

4.3 Prostate Zonal Anatomy

The prostate zonal anatomy proposed by McNeal (McNeal et al. 1988) is of utmost importance to the interpretation of MRI of prostate pathologies, particularly in prostate cancer (Skilinda et al. 2019). In this classification, the prostate is divided into four zones: peripheral zone, central zone, transition zone, and anterior fibromuscular zone (Fig. 4.4).

The glandular tissue of the prostate in adults represents two fused glands in a single structure. These two parts are called the central zone and peripheral zone (McNeal et al. 1988). The central zone consists of a portion of the glandular tissue surrounding the ejaculatory ducts. It has an apex, located near the seminal colliculus (verumontanum), and a base, located above and behind the bladder neck (Fig. 4.4).

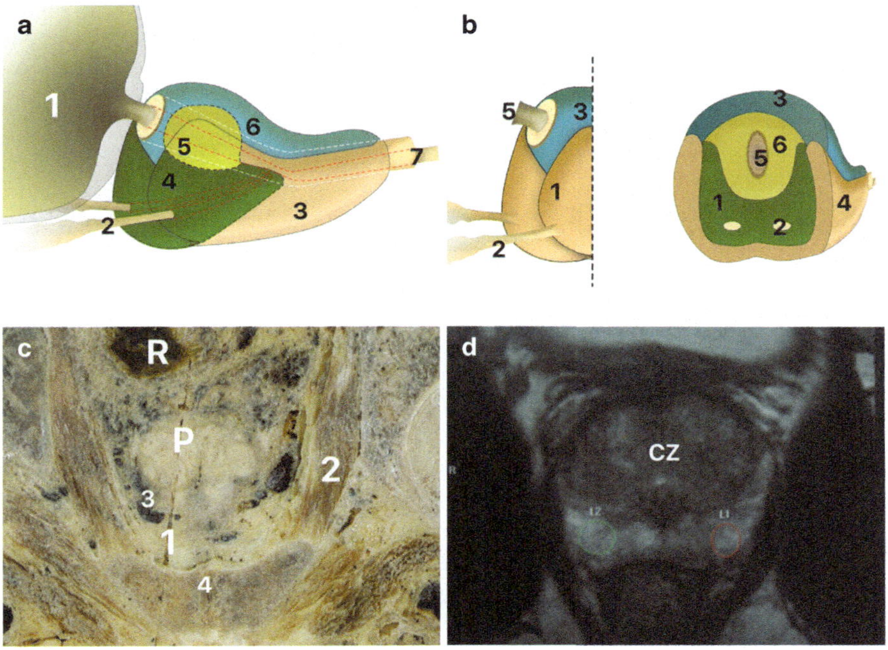

Fig. 4.4 Zonal classification of the prostate. (**a**) Schematic drawing showing the zonal anatomy of the prostate according to McNeal classification (McNeal et al. 1988). 1, bladder; 2, ejaculatory duct; 3, peripheric zone; 4, central zone; 5, transition zone; 6, anterior zone (anterior fibromuscular stroma); 7, urethra. (**b**) Schematic drawing showing a frontal section of the prostate at the level of the ejaculatory ducts. 1, central zone; 2, ejaculatory duct; 3, anterior zone; 4, peripheric zone; 5, urethra; 6, transition zone. (**c**) Transverse section of a male pelvis at the level of the pubic symphysis (4) showing the relationships of the prostate (P). R, rectum; 1, Retzius space; 2, internal obturator muscle; 3, periprostatic venous plexus. (**d**) Magnetic resonance image of a male pelvis showing the prostatic zones, peripheric zone (L1 and L2), and central zone (CZ)

The central zone accounts for about 20% of the total mass of the prostatic glandular tissue.

The peripheral zone is the largest region of the prostate, making up about 70% of the total glandular mass. It is represented by a pair of ducts that originate in the posterolateral recess of the urethra wall and radiate laterally (Fig. 4.4). Together, the central and peripheral zones account for 90% to 95% of the prostatic glandular tissue (McNeal et al. 1988).

The transition zone is the innermost section of the prostate and forms 5% of the prostate. The most distal region of the periurethral sphincter contains more complex ducts and a larger acinar system. This relationship between the tissues of different types and origins can be important to understand benign prostatic pathology. Finally, a considerable region of the prostate, called the anterior zone (anterior fibromuscular stroma), is entirely nonglandular, instead consisting mainly of smooth muscle fibers. The anterior fibromuscular stroma is connected to the inferior edge of pubic symphysis by the puboprostatic ligament. The anterior zone does not play an important role in prostate function, but its location and extension increase the difficulty of visualizing and studying the glandular tissue of the anterolateral region of the prostate (McNeal et al. 1988).

4.4 Prostatic Urethra

The prostatic urethra is 3–4 cm long and is located between the prostate apex and penile bulb. It descends from the base to the apex of the prostate, nearly vertically forming anteriorly a concave curve, and is covered by a cylinder with vertical orientation called the external striated urethral sphincter (Oerlich 1980; Brooks et al. 1998). In the posterior aspect of the urethra lies the urethral crest with the seminal colliculus, where ejaculatory ducts open (Fig. 4.5).

Near the bladder neck, the fibers decrease and eventually disappear, but a few can be seen mixed with the smooth muscle tissue on the lateral face of the vesical trigone. This smooth muscle forms the internal urethral sphincter as it emerges from the anterior wall of the bladder and gradually decreases over the anterior face of the prostate. Some of its fibers probably penetrate the pubovesical or puboprostatic ligament (Manley 1966).

4.5 Seminal Vesicles and Ejaculatory Ducts

The seminal vesicles are a pair of glands attached to the male genital system, responsible for producing part of the seminal fluid. Each seminal vesicle has an elongated pyriform shape, with a broadened upper end and lower end or neck joining the vas deferens (Fig. 4.6). It contains several internal cavities, and its wall is mainly formed (about 80%) by smooth muscle tissue (Nguyen et al. 1996).

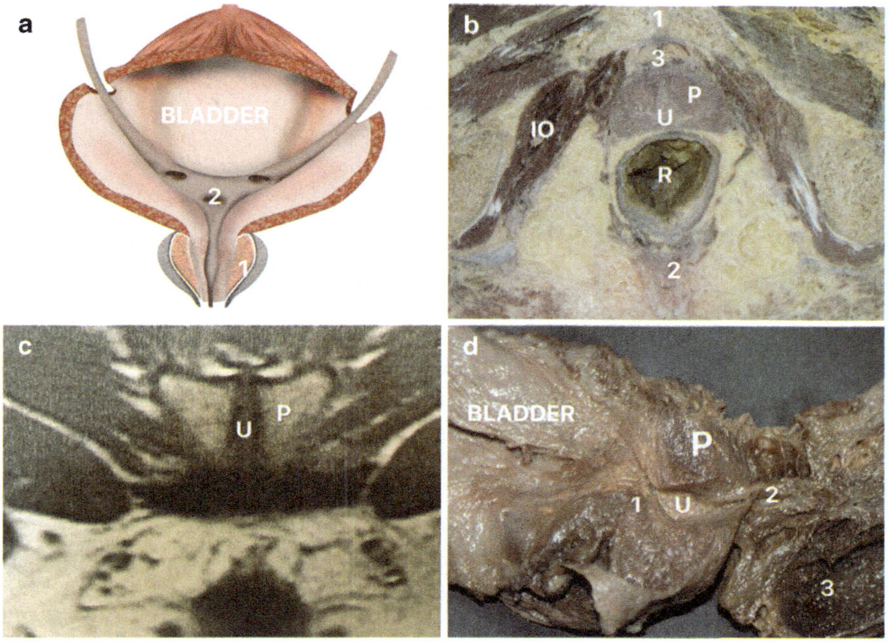

Fig. 4.5 (**a**) Schematic drawing showing the relationships between the prostate (1) and the bladder; 2, bladder trigone. (**b**) Transverse section of a male pelvis at the level of the pubic symphysis (1) showing the relationships of the prostate (P). R, rectum; 2, sacrum; 3, Retzius space; IO, internal obturator muscle. (**c**) Magnetic resonance image of a male pelvis showing the prostate (P) and the prostatic urethra (U). (**d**) The figure shows a fixed part of a male corpus showing the position of the prostatic urethra (U); P, prostate; 1, seminal colliculus; 2, membranous urethra; 3, penile bulb

Knowledge of the seminal glands is important for surgery and MRI interpretation in prostate cancer. Anteriorly, each seminal vesicle is related to the portion of the bladder corresponding to the vesical trigone. It is possible to analyze the seminal vesicles by means of a digital rectal exam and transrectal ultrasound. Laterally, the seminal vesicles are related to the periprostatic venous plexus (Fig. 4.6).

Each ejaculatory duct is formed by the union at an acute angle of the seminal vesicle and the vas deferens. After a path of 15–20 mm in adults, the ejaculatory ducts open into the prostatic urethra in a small ostium located in the rear part of the seminal colliculus, one on the right and one on the left of the prostatic utricle. The ejaculatory duct can be divided into three segments: proximal, median, and distal (Nguyen et al. 1996). The proximal segment is located at the junction with the seminal vesicle (Fig. 4.6). The duct is the right continuation of the seminal vesicle, while the ampoule of the vas deferens enters medially at an acute angle. The intraprostatic portion or median segment extends on the outside of the prostate surface for 10–15 mm. The distal segment, located in the central region of the prostate, opens into the seminal colliculus (Weingartner et al. 1996).

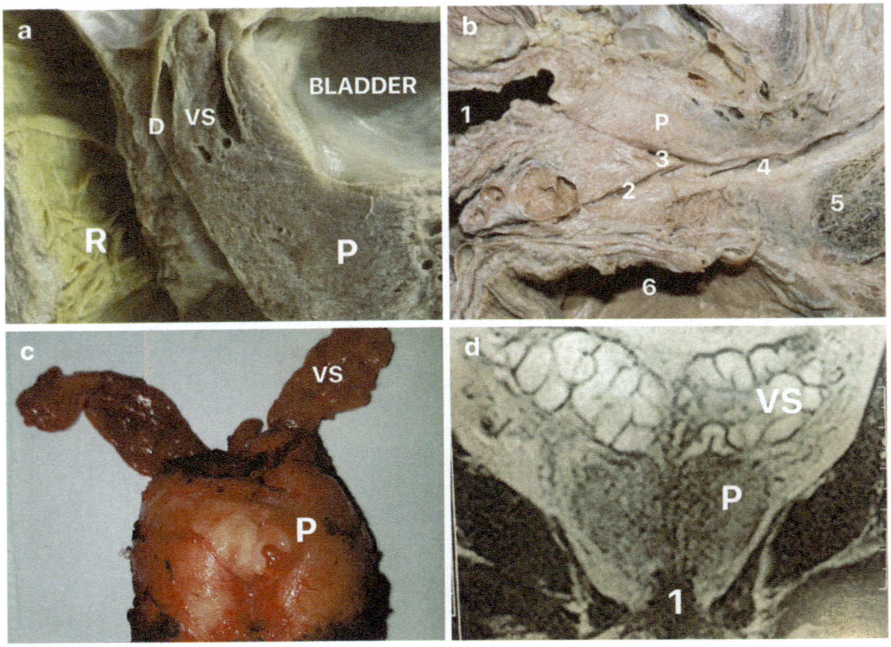

Fig. 4.6 (**a**) Sagittal section of a male pelvis from a fixed corpse indicating the relationships between the prostate (P) and seminal vesicle (VS); R, rectum; D, Denonvilliers fascia. (**b**) Sagittal section of a male pelvis from a fixed corpse showing the ejaculatory duct (2); 1, bladder; 3, verumontanum; 4, membranous urethra; 5, penile bulb; 6, rectum. (**c**) The figure shows a surgical piece after a radical prostatectomy showing the prostate (P) and the seminal vesicles (VS). (**d**) Magnetic resonance image of a male pelvis showing the seminal vesicles (VS), prostate (P), and urethral sphincter (1)

4.6 Scrotum

Knowledge of scrotal sectional anatomy is very important to the radiological interpretation of testicular pathologies, especially testicular torsion and testicular cancer. The scrotal sack is formed by a skin fold in the perineal region and contains the testes, epididymis, and elements of the spermatic funiculus (Netter 1978). The scrotal sack is divided into two independent compartments by a median raphe (Fig. 4.7). Infectious pathologies or the buildup of liquids in one of the compartments usually does not disseminate to the other side due to the presence of this anatomical barrier (Parker and Robison 1971).

The testicle is the most important element of the scrotum. It has two extremities, one superior and one inferior; two lateral edges; and two surfaces, one anterior and one posterior, which are covered by the epididymis. A testicular appendage is located at the upper end of each testis. The testis is enveloped by a thick tunica with a large quantity of dense connective tissue, the tunica albuginea, which sends septa to the interior of the testis, dividing it into lobes, where the convoluted seminiferous

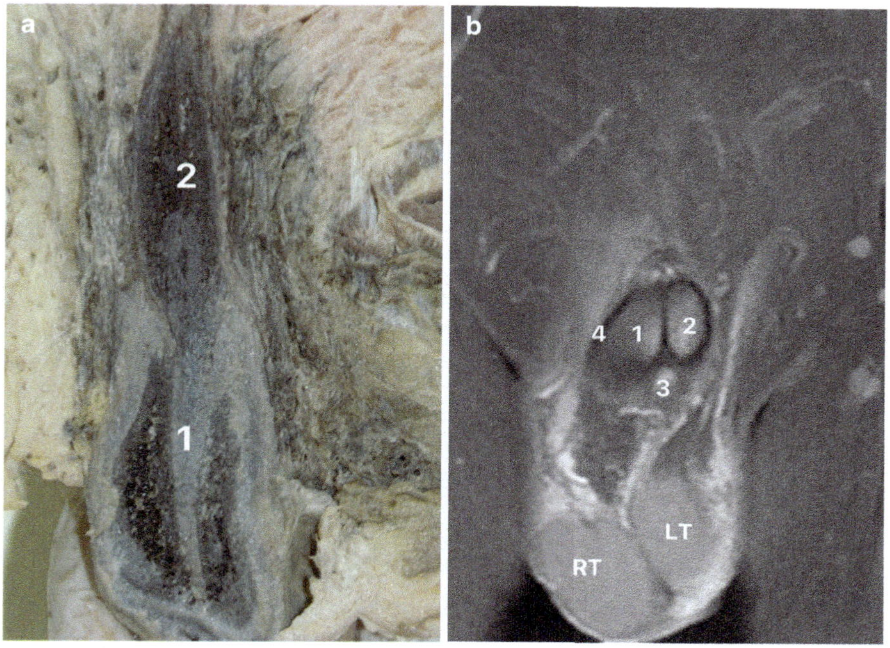

Fig. 4.7 (**a**) Frontal section of a male pelvis from a fixed corpse showing the scrotal raphe (1) and the penile bulb (2). (**b**) Magnetic resonance image of a male pelvis showing the penis and the scrotum; RT, right testis; LT, left testis; 1, right cavernous body; 2, left cavernous body; 3, penile urethra; 4, tunica albuginea

tubules are located (Fig. 4.8). These tubules reach the mediastinum testis, where they become straight and give rise to the testicular network from where the efferent tubules depart, which establish the communication between the testis and head of the epididymis (Parker and Robison 1971).

The epididymis is located on the posterolateral face of the testis. It has dilated upper portion, called head; a central portion, the body; and a lower tapered end, the tail. The head of the epididymis is directly connected to the testis cranial pole by the efferent ducts. The tail is linked to the testis inferior pole by areolar tissue and a reflection of the tunica vaginalis. A recession of the tunica vaginalis, the sinus of the epididymis or mesorchium, is found between the body of the vas deferens and the lateral face of the testis (Netter 1978). The vas deferens is a continuation of the epididymal duct that transports the spermatozoids from the epididymis to the ejaculatory duct. The vas deferens starts in the tail of the epididymis, where it is sinuous, and ascends on the medial side of the epididymis where it is surrounded by the pampiniform plexus, becoming part of the spermatic cord. It continues in an upward direction to the external inguinal ring, and in this region, it is easily palpable. It is the rearmost element with the hardest consistency of the spermatic cord. After passing through the inguinal canal, it curves around the inferior epigastric artery and

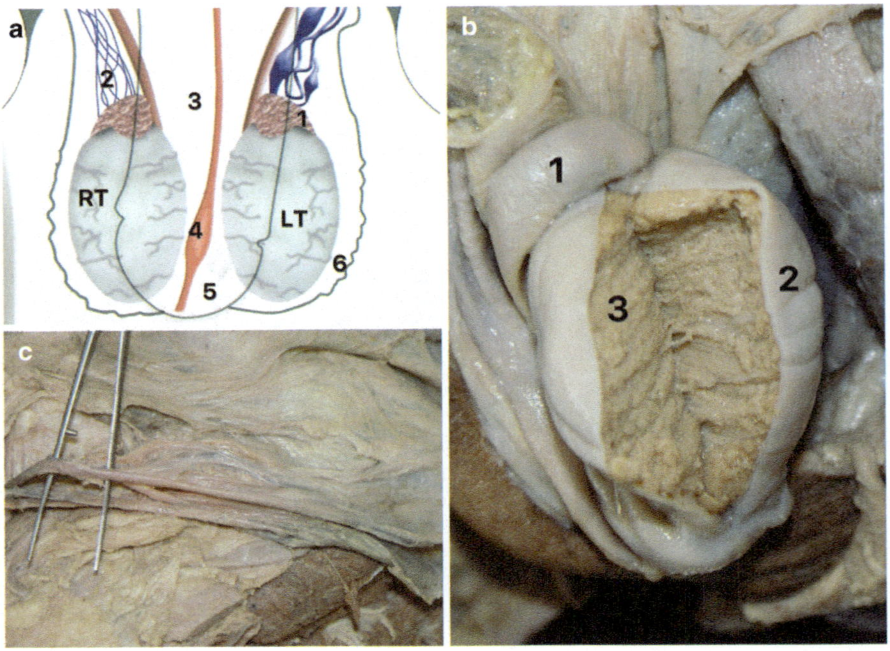

Fig. 4.8 (**a**) Schematic drawing showing the testicular relations; RT, right testis; LT, left testis; 1, epididymis head; 2, testicular venous plexus; 3, shadow showing the penis body; 4, navicular fossa of urethra; 5, shadow showing the penis gland; 6, shadow showing the scrotum. (**b**) In the same fixed corpse, the epididymis (1) and testis were dissected and the albuginea tunica (2) was opened to show the testicular parenchyma (3). (**c**) The figure shows the right inguinal region of a fixed corpse showing the testicular vessels and deferens duct

ascends anteriorly to the external iliac artery, returning posteriorly and inferiorly and penetrating the pelvis (Fig. 4.8). It crosses the medial face of the ureter, meets the posterior face of the bladder, and extends in the inferior and medial direction on the medial face of the seminal vesicle. In this region, the duct is dilated and sinuous and is called ampoule, joining the seminal vesicle duct to form the ejaculatory duct.

4.7 Penis

Knowledge of the sectional anatomy of the penis is very important to the interpretation of radiological exams, particularly the MRI. MRI is very useful in diagnosing penile pathologies such as penile trauma and penile cancer (Tu et al. 2020). The penis is divided into two portions: the root, located in the surface space of the perineum, which is responsible for the attachment and stability of the penis; and the free portion, which composes the largest part of the organ and is composed of three erectile structures, two corpora cavernosa and a corpus spongiosum, which presents a distal expansion covered by skin, the glans penis (Netter 1978) (Fig. 4.9).

The root of the penis is formed by a dilated central extremity, the bulb, and two lateral parts, the branches or pillars of the penis (Fig. 4.9). The bulb is located in the interval between the two branches and is attached to the inferior surface of the lower face of the urogenital diaphragm, from where it continues anteriorly with the corpus spongiosum and is crossed in its central portion by the bulbar urethra and surrounded by the bulbospongiosus muscle (Netter 1978).

The free portion of the penis is divided into the body and glans, the latter of which is covered by a skin fold called the foreskin or prepuce. The upper part of the penile body is also called the dorsal region, and the lower part, through which the urethra runs, is called the ventral region. The glans is the dilated distal portion of the corpus spongiosum. It has a slight elevation that separates it from the body of the penis, called corona, and is separated from the prepuce by the balanopreputial fold. Prepuce has a vascularized extension from the mucosa that is attached to the ventral portion of the glans, denominated frenulum of prepuce. The tip of the glans is the location of the external urethral meatus. This region is also the location of the navicular fossa, the second most dilated region of the urethra, only behind the prostatic

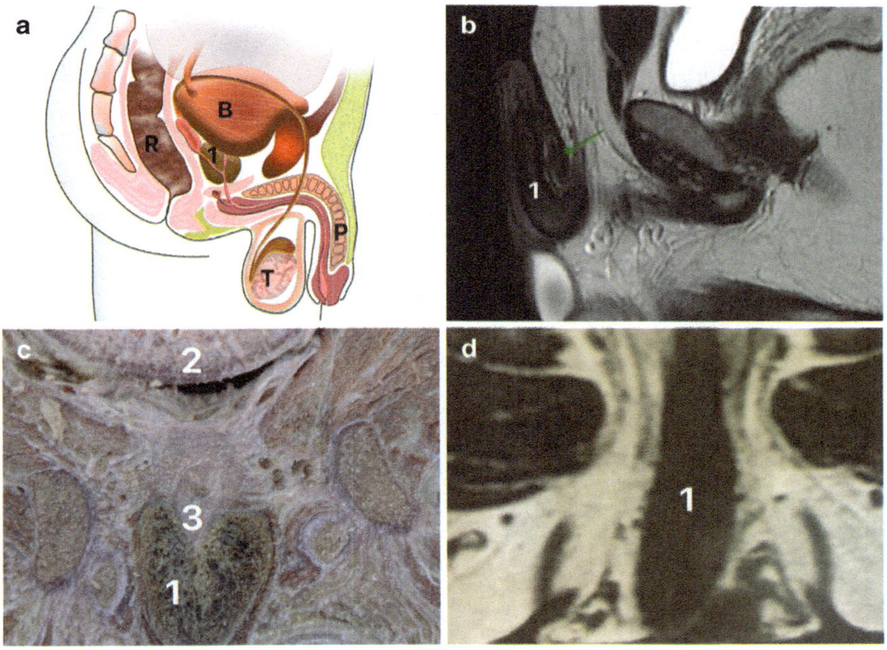

Fig. 4.9 (**a**) Schematic drawing showing the penis (P) relationships; B, bladder; R, rectum; T, testis; 1, prostate. (**b**) Magnetic resonance image of a patient with penile trauma with dorsal vein lesion (green arrow); 1, cavernous body. (**c**) Frontal section of a fixed male corpse behind the pubic symphysis showing the penile bulb (1); 2, bladder; 3, bulbar urethra. (**d**) Magnetic resonance image of a male pelvis showing the penile bulb (1)

urethra (Netter 1978). The navicular fossa is better visualized when making a sagittal cut of the glans (Fig. 4.10).

The body of the penis is formed by three erectile structures that become engorged with blood during erection, two corpora cavernosa and a corpus spongiosum. The two corpora cavernosa are located on the dorsal region of the penis, i.e., the postero-superior region, when the penis is erect. The corpus spongiosum is crossed by the urethra and is in the ventral region of the penis, i.e., the region which is in contact with the perineal region when the penis is flaccid. The corpus spongiosum continues in a dilated distal region to the glans penis.

A transversal cut of the body of the penis (Fig. 4.10) enables observing that the erectile bodies are surrounded by some series of structures. Just below the skin is the subcutaneous tissue, called the superficial fascia of the penis. This superficial fascia contains a small quantity of smooth muscle fibers and is virtually bereft of fat. It continues with the tunica dartos of the scrotal sack and with the fascia of the perineum. Below the fascia of the penis, there is a continuation of the deep perineal fascia, the

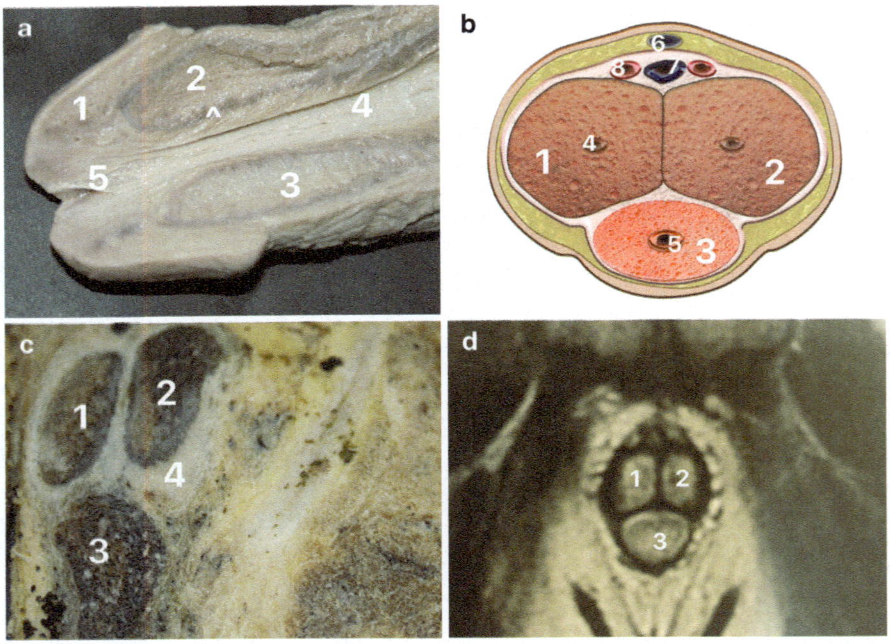

Fig. 4.10 (**a**) The figure shows a penile sagittal section; we can observe the glans (1), the right (2) and left (3) cavernous bodies, penile urethra (4), navicular fossa (5), and the tunica albuginea (arrowhead). (**b**) Schematic drawing of a transverse section of the penile body showing the right (1) and left (2) cavernous bodies, the spongiosum body (3), the cavernous arteries (4), penile urethra (5), superficial dorsal penile vein (6), deeper dorsal penile vein (7), and dorsal penile artery (8). (**c**) The figure shows a transverse section of the penile body of a fixed corpse; we can observe the right (1) and left (2) cavernous bodies, the spongiosum body (3), and the albuginea tunica (4). (**d**) Magnetic resonance image of penis showing the right (1) and left (2) cavernous bodies and the spongiosum body (3)

deep fascia of the penis, or Buck's fascia. It is a strong and membranous fascia that surrounds the two corpora cavernosa and the corpus spongiosum (Netter 1978).

The deep fascia of the penis is surrounded by a dense system of fibers, called the tunica albuginea. The portion of the tunica albuginea that surrounds the corpora cavernosa is thinner than the part that surrounds the corpus spongiosum (Fig. 4.10). Numerous trabeculae originate from the tunica albuginea, dividing the interior of the erectile bodies into cavernous spaces (Goldstein et al. 1985).

4.8 Vascularization of Male Pelvis

The most important artery of the male pelvis is the internal iliac artery (hypogastric artery). The prostate is irrigated by the inferior vesical artery and medial rectal artery, branching from the anterior trunk of the internal iliac artery. The inferior vesical artery is the most important, and when it approaches the gland, it splits into two main branches: urethral artery and capsule artery. The urethral artery penetrates the gland through the prostate-vesical junction, posteriorly, and connects to the bladder neck, being responsible for a large part of the vascularization in cases of benign prostatic hyperplasia. When the adenoma is resected, the main bleeding points are on the bladder neck in the 4 and 8 o'clock positions (Brooks 2002).

The periprostatic venous plexus is derived from the deep dorsal vein of the penis, which passes between the pubic arch and the striated urethral sphincter, entering the pelvis, where it divides into three branches: the central superficial branch and two lateral plexus branches. The superficial branch is located between the puboprostatic ligaments and drains the retropubic fat, the front wall of the bladder and the rear portion of the prostate (Fig. 4.3). The lateral plexuses are located beside the prostate and receive drainage from the rectum in communication with the vesical plexus. There are three to five veins that originate from the lateral plexus and drain to the internal iliac vessel. These veins communicate with the emissary veins of the pelvic bones and with the vertebral plexus, thought to be involved in the dissemination of prostate tumors (Brooks 2002). Variations are frequent in the location and distribution of the periprostatic veins (Myers 1991).

Each testis is irrigated by three arteries: testicular artery, a branch of the right aorta; deferential artery; and cremasteric artery, a branch of the internal iliac artery. These three arteries penetrate the organ in the mediastinum testis region, where they provide ample communication. The testes are drained by the pampiniform venous plexus, which in the region of the deep inguinal ring originates the testicular veins. The left testicular vein opens into the left renal vein, and the right testicular vein opens directly into the inferior cava vein (Netter 1978).

The penis is irrigated by two internal pudendal arteries, branches of internal iliac (hypogastric) artery. After its various perineal branches, the pudendal arteries combine to form the so-called common penile artery, which divides into three branches: the bulbourethral artery, the dorsal penile artery, and the cavernosal artery (Fig. 4.11). The cavernosal artery is located inside the corpus cavernosum, the bulbourethral artery is responsible for irrigating the corpus spongiosum and urethra, and the

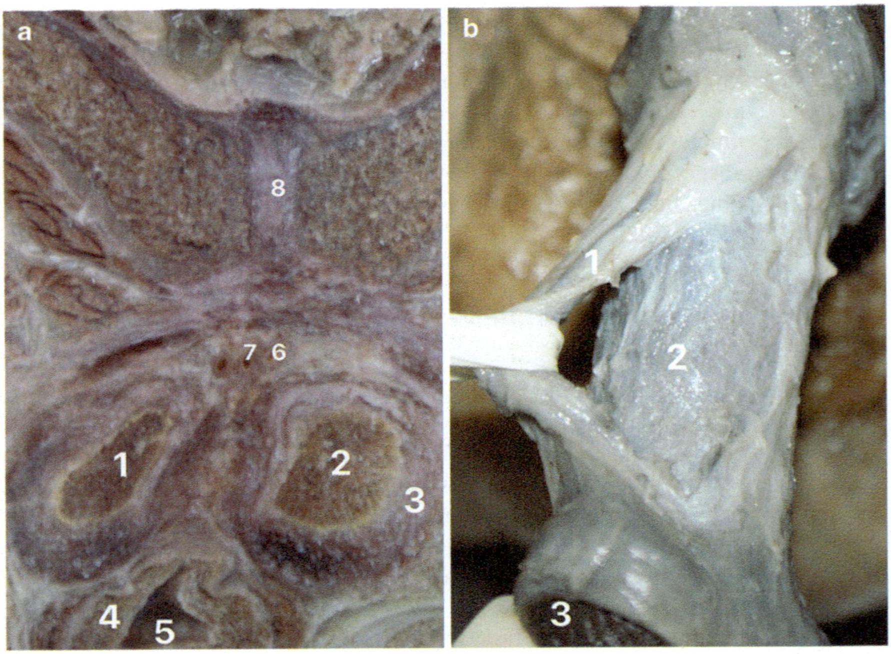

Fig. 4.11 (**a**) Transverse section of a male fixed corpse at the level of pubic symphysis (8) showing the transition of penile body and penile root; 1, right cavernous body; 2, left cavernous body; 3, tunica albuginea; 4, spongiosum body; 5, penile urethra; 6, dorsal penile artery; 7, deeper dorsal penile vein. (**b**) The figure shows a dissection of the penile body showing the neurovascular structures situated at the dorsal region of the penis (1), the cavernous body (2), and the glans (3)

dorsal penile artery is located between the tunica albuginea and Buck's fascia. These penile arteries branch out, giving rise to smaller arteries that run to the cavernous spaces, the helicine arteries (Netter 1978).

The venous drainage of the penile erectile tissue originates in small venules that start in the perisinusoidal spaces located below the tunica albuginea. These venules connect to the emissary veins, which cross the tunica albuginea and open into the circumflex veins, located beside the erectile bodies. The circumflex veins drain into the deep penile dorsal vein, which in turn opens into the periprostatic venous plexus. The circumflex veins have one or more valves. The skin and subcutaneous tissue of the penis are drained by the superficial penile dorsal vein, which generally empties into the greater saphenous vein.

4.9 Lymphatic Drainage of Male Pelvis

The lymphatic drainage of the pelvis is performed mainly by the internal iliac lymph nodes. These lymph nodes have three main chains: pre-sacral, obturator, and internal pudendal lymph nodes. The lymphatic drainage of the prostate goes to the internal iliac lymph nodes, principally to obturator nodes (Weingartner et al. 1996).

The lymphatic drainage of the scrotal sack is performed by the superficial inguinal lymph nodes. The right testis drains to the retroperitoneal lymph nodes, located along the renal pedicle and inferior cava vein, as well as between the inferior cava vein and the aorta. The left testis drains to the lymph nodes located along the left renal hilum and aorta artery. One of the main routes for dissemination of testicular tumors is the lymphatic system. Knowledge of the lymphatic drainage of the testicles and the scrotal sack is important for proper diagnosis and therapy of these tumors.

The penile skin and prepuce drain to the superficial lymph nodes in the inguinal region, located above the fascia lata. The glans and the rest of the penis drain to the deep inguinal lymph nodes, located under the fascia lata, and to the external iliac lymph nodes. The so-called sentinel lymph node of Cabanas is of particular importance in the dissemination of penile tumors. It is located medially across from the great saphenous vein and is generally the first place afflicted in the lymphoid dissemination of penile tumors (Netter 1978).

4.10 Innervation of the Male Pelvis

The innervation of the pelvis is derived from the sacral and coccygeal segments of the spinal cord and the pelvic portion of the autonomic nervous system. The bladder receives autonomic innervation of the bladder and prostatic plexus. The striated sphincter of the urethra is innervated by the pudendal nerve.

The pudendal nerve originates from the sacral plexus and is responsible for most of the innervation of the perineum (Netter 1978). It contains sensitive motor fibers and preganglionic sympathetic fibers (Netter 1978). It crosses the greater sciatic foramen, penetrating the pelvis through the lesser sciatic foramen, together with the internal pudendal vessels, crossing the pudendal canal (Alcock's canal). In the lateral wall of the ischiorectal fossa, it divides, originating the inferior rectal nerve, the perineal nerve, and the dorsal nerve of the penis (clitoris) (Fig. 4.12). The pudendal nerve is responsible for the somatic innervation of the skeletal muscles that make up the external sphincter of the urethra (Hinata et al. 2013).

The autonomic innervation of the bladder and urethra comes from the bladder and prostate plexuses, which originate from the lower hypogastric plexus. The superior hypogastric plexus, below the aortic bifurcation, divides into two lower right and left hypogastric nerves (Netter 1978) (Fig. 4.12). The inferior hypogastric nerve crosses the upper pelvis and joins the sacral splanchnic nerves, giving rise to the inferior hypogastric plexus (Hinata et al. 2013). The sympathetic fibers are derived from the T11 to L2 medullary segments and go to the superior hypogastric plexus. The parasympathetic fibers come from the S2, S3, and S4 medullary segments and reach the inferior hypogastric plexus through the pelvic splanchnic nerves.

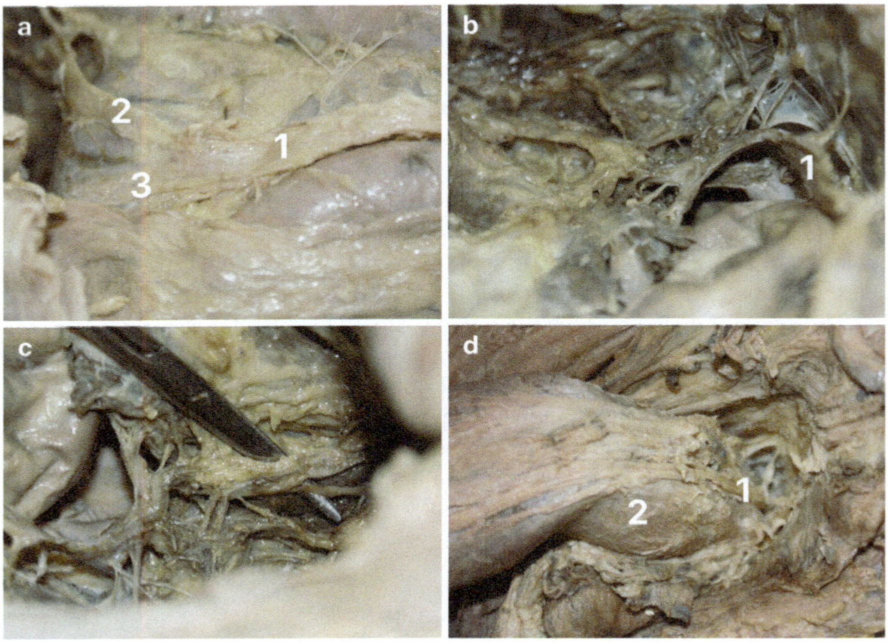

Fig. 4.12 Innervation of a male pelvis. (**a**) Dissection of a male pelvis at the level of the first sacral vertebra showing the superior hypogastric plexus (1) and the right (2) and the left (2) hypogastric nerves. (**b**) The right hypogastric nerve head for pelvic cavity receives the pelvic splanchnic nerves and originates the inferior hypogastric plexus. (**c**) The scissors show some branches of the inferior hypogastric plexus. (**d**) The figure shows a penile dissection with the cavernous nerves (1) that originates the dorsal nerve of the penis; 2, cavernous body

The innervation of the prostate originates from the inferior hypogastric plexus, which in turn originates from the union of the hypogastric nerve (branch of the superior hypogastric plexus) with the pelvic splanchnic nerves (parasympathetic branches) (Weingartner et al. 1996). One of the main branches of the inferior hypogastric plexus is the cavernous nerve (neurovascular bundle), responsible for erection. A close relationship exists between the cavernous nerve and the lateral surface of the prostate.

The penis is innervated by the dorsal penile nerves, which are branches of the pudendum nerve, in turn innervating the skin and mainly the glans. The deep branches of the perineal nerves, which enter the bulb and mainly innervate the urethra and cavernous penile nerves, in turn are branches of the inferior hypogastric plexus, responsible for the autonomic innervation of the penis, in particular the penile erectile bodies (Ophoven and Roth 1997) (Fig. 4.12).

References

Brooks JD. Anatomy of the lower urinary tract and male genitalia. In: Campbell's urology. 8th ed. New York: Saunders; 2002. p. 41.

Brooks JD, Chao WM, Kerr J. Male pelvic anatomy reconstructed from the visible human data set. J Urol. 1998;159:868–72.

Goldstein AMB, Meehan JP, Morrow JW, Buckley PA, Rogers FA. The fibrous skeleton of the corpora cavernosa and its probable function in the mechanism of erection. Brit J Urol. 1985;57:574–8.

Hinata N, Sejima T, Takenaka A. Progress in pelvic anatomy from the viewpoint of radical prostatectomy. Int J Urol. 2013;20:260–70.

Julio Junior HR, Costa SF, Costa SW, Sampaio FJB, Favorito LA. Structural study of endopelvic fascia in prostates of different weights. Anatomic study applied to radical prostatectomy. Acta Cir Bras. 2015;30:301–5.

Manley CB. The striated muscle of the prostate. J Urol. 1966;95:234–40.

McNeal JE, Redwine EA, Freiha FS, Stamey TA. Zonal distribution of prostatic adenocarcinoma: correlation with histologic pattern and direction of the spread. Am J Surg Path. 1988;12:897–907.

Myers RP. Anatomical variation of the superficial preprostatic veins with respect to radical retropubic prostatectomy. J Urol. 1991;145:992–3.

Netter F. Reproductive System. Vol 2. Ciba, New Jersey, 1978.

Nguyen HT, Etzell J, Turek PJ. Normal human ejaculatory duct anatomy: a study of cadaveric and surgical specimens. J Urol. 1996;155:1639–42.

Oerlich TM. The urethral sphincter muscle in the male. Am J Anat. 1980;158:229–46.

Ophoven A, Roth S. The anatomy and embryological origins of the fascia of Denoviliers: A medico-historical debate. J Urol, 1997;157:3–9.

Parker RM, Robison JR. Anatomy and diagnosis of torsion of the testicle. J Urol. 1971;106:243–7.

Patel VR, Schatloff O, Chauhan S, et al. The role of the prostatic vasculature as a landmark for nerve sparing during robot-assisted radical prostatectomy. Eur Urol. 2012;61:571–6.

Raychaudhuri B, Cahill D. Pelvic fasciae in urology. Ann R Coll Srug Engl. 2008;90:633–7.

Skilinda K, Fracksek M, Mruk B, Walecki J. Normal 3T MR anatomy of the prostate gland and surrounding structures. Adv Med, Article ID 3040859. 2019:9.

Takenaka A, Hara R, Soga H, Murakami G, Fujisawa M. A novel technique for approaching the endopelvic fascia in retropubic radical prostatectomy, based on an anatomical study of fixed and fresh cadavers. BJU Int. 2005;95:766–71.

Takenaka A, Tewari AK, Leung RA, Bigelow K, El Tabey N, Murakami G, Fujisawa M. Preservation of the puboprostatic collar and puboperineoplasty for early recovery of urinary continence after robotic prostatectomy: anatomic basis and preliminary outcomes. Eur Urol. 2007;51:433–40.

Tu LH, Spektor M, Ferrante M, Mathur M. MRI of the penis: indications, anatomy, and pathology. Curr Probl Diagn Radio. 2020;49(1):54–63.

Vora AA, Dajani D, Lynch JH, Kowalczyk KJ. Anatomic and technical considerations for optimizing recovery of urinary function during robotic-assisted radical prostatectomy. Curr Opin Urol. 2013;23:78–87.

Walz J, Burnett AL, Costello AJ, et al. A critical analysis of the current knowledge of surgical anatomy related to optimization of cancer control and preservation of continence and erection in candidates for radical prostatectomy. Eur Urol. 2010;57:179–92.

Weingartner K, Ramaswamy A, Bittinger A, Gerharz EW, et al. Anatomical basis for pelvic lymphadenectomy in prostate cancer: results of an autopsy study and implications for the clinic. J Urol. 1996;156:1969–71.

Chapter 5
Sectional Anatomy of the Female Pelvis

Luciano Alves Favorito, Natasha T. Logsdon, and Francisco J. B. Sampaio

5.1 Introduction

Knowledge of the sectional anatomy of the female pelvis (pelvic floor and female organs) is essential for understanding the mechanisms of urinary continence and pathologies of the female pelvis and for the treatment of incontinence. So, we will review the main structures involved in this process, discussing the role of each one of them and showing the sectional anatomy of these structures and the relationships with the image exams.

L. A. Favorito (✉)
Urogenital Research Unit, Rio de Janeiro State University, Rio de Janeiro, RJ, Brazil

IDOMED-UNESA, Rio de Janeiro, RJ, Brazil

The National Council for Scientific and Technological Development (CNPq), Lago Sul, Federal District, Brazil

The Rio de Janeiro State Research Foundation (FAPERJ), Rio de Janeiro, RJ, Brazil

N. T. Logsdon
University Center Geraldo di Biasi, Rio de Janeiro, RJ, Brazil

Urogenital Research Unit, Rio de Janeiro State University, Rio de Janeiro, RJ, Brazil

F. J. B. Sampaio
Urogenital Research Unit, Rio de Janeiro State University, Rio de Janeiro, RJ, Brazil

The National Council for Scientific and Technological Development (CNPq), Lago Sul, Federal District, Brazil

The Rio de Janeiro State Research Foundation (FAPERJ), Rio de Janeiro, RJ, Brazil
e-mail: sampaio@uerj.br

L. A. Favorito, N. T. Logsdon (eds.), *Atlas of Sectional Anatomy*,
https://doi.org/10.1007/978-3-030-91688-6_5

5.2 Pelvic Floor

The pelvic floor and the set of structures that support abdominal and pelvic viscera are of great importance. The pelvic floor therefore includes the peritoneum, the pelvic diaphragm, the urogenital diaphragm, and the structures located between the peritoneum and these diaphragms (Williams et al. 1995). The pelvic floor plays an important role in urinary continence and pelvic-organ support in females (Wu et al. 2020). The pelvic diaphragm is composed of the levator ani muscle and coccygeal muscle, in addition to the fasciae that cover them superiorly and inferiorly.

5.2.1 Levator Ani Muscle (LAM)

The LAM forms most of the pelvic diaphragm, above which the pelvic viscera are suspended (Fig. 5.1). The LAM is funnel-shaped, with attachments to the pubic bone anteriorly, and the internal obturator muscle and ischial spine laterally (House et al. 2009; Wu et al. 2020). Each levator ani muscle has a linear origin, from the dorsal body of the pubis, the pelvic fascia that covers the internal obturator muscle and the ischial spine, and divides the pelvic cavity of the ischiorectal fossae (Fig. 5.1). Its fibers are divided into groups with different insertions. a) Anterior fibers: they form the puborectalis muscle, in which its pre-rectal fibers form a loop around the vagina and insert itself in the tendinous center of the perineum. This is the portion of the muscle that is most prone to ruptures during childbirth and the one that must be sectioned in episiotomies. b) Intermediate fibers: they compose the pubococcygeus muscle. It forms a loop around the junction of the rectum with the anal canal, inserting itself in the anococcygeal ligament. c) Posterior fibers: they compose the iliococcygeus muscle, which inserts in the anococcygeal ligament and the coccyx (Fig. 5.1) (Rouviére 1968; Netter 1978).

The LAM forms a ring that supports and holds the pelvic viscera in position, offering resistance to increases in intra-abdominal pressure during the efforts of horny and expulsion. These muscles have an important sphincter action on the anorectal junction, on the urethra under stressful conditions, and on the vagina. Its innervation is derived from the III and IV sacral nerves, and its anterior portion has innervation of the perineal branch of the pudendal nerve (Chiara 1959) (Fig. 5.2). Between the muscle bundles of the levator ani, there are slits that create communications between the pelvic subperitoneal connective tissue, the ischiorectal fossa, and the gluteal and obturator regions (Williams et al. 1995) (Fig. 5.2).

The coccygeus muscle is a small muscle that originates in the ischial spine and inserts in the inferior end of the sacrum and in the superior part of the coccyx (Fig. 5.2). It assists the levator ani to support pelvic viscera. Its innervation derives from the IV sacral nerve (perineal branch).

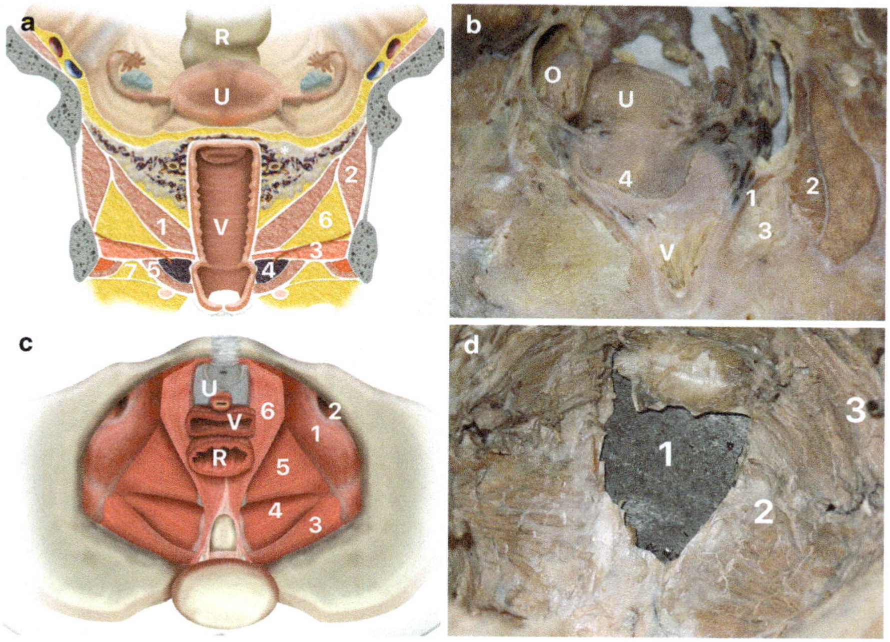

Fig. 5.1 The figure shows some aspects of the female pelvic floor. (**a**) Schematic drawing of a frontal section showing some details of the female pelvic floor. R, rectum; U, uterus; V, vagina; 1, levator anus muscle (LAM); 2, internal obturator muscle; 3, urogenital diaphragm; 4, vestibule bulb; 5, bulbospongiosus muscle; 6, ischiorectal fossa; 7, superficial perineal space. (**b**) The figure shows a frontal section of a fixed corpus showing some details of the female pelvic floor. O, ovarium; U, uterus; V, vagina; 1, levator anus muscle (LAM); 2, internal obturator muscle; 3, ischiorectal fossa; 4, uterine cervix. (**c**) Schematic drawing of a superior view of the female pelvic floor showing some details of pelvic diaphragm and LAM. R, rectum; U, uterus; V, vagina; 1, internal obturator muscle; 2, obturator canal; 3, piriform muscle; 4, coccygeal muscle; 5, iliococcygeus muscle; 6, pubococcygeus muscle. (**d**) Superior view of a fixed corpus showing the LAM (2) and this relationship with the internal obturator muscle (3); the female viscera pelvic were suppressed, and the pelvic diaphragm stays around the visceral hiatus (1)

5.2.2 Female Anterior Perineum and Urogenital Diaphragm

The female anterior perineum is constituted by skin, subcutaneous mesh, superficial perineal space and its content, urogenital diaphragm, and deep space of the perineum (Williams et al. 1995). The superficial perineal space (Figs. 5.1 and 5.3) is limited below by the membranous layer of the superficial fascia and above by the urogenital diaphragm. In this space, there are the bulbs of the vestibule, the branches of the clitoris, the greater vestibular glands (or Bartholin glands), and the bulky-spongy, ischiocavernosus, and superficial transverse muscles of the perineum, in addition to loose connective tissue with vessels and nerves.

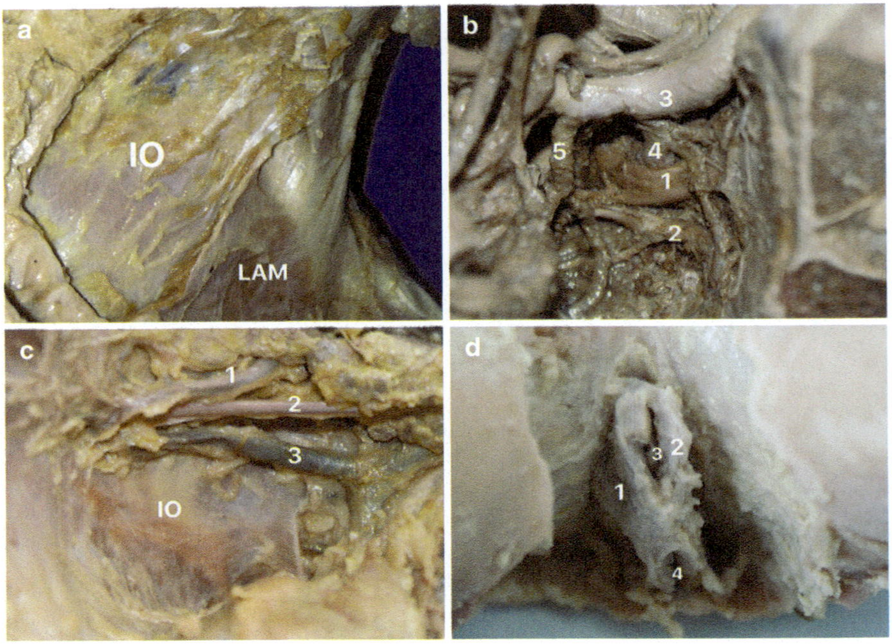

Fig. 5.2 The figure shows some aspects of the female pelvic floor and perineum. (**a**) The figure shows the relationships between the levator ani muscle (LAM) with the internal obturator muscle (IO); the LAM inserts in the IO fascia. (**b**) The figure shows some arteries and pelvic nerves; 1, hypogastric artery; 2, nerve; 3, nerve; 4, artery; 5, uterine artery. (**c**) The figure shows the internal obturator muscle (IO) and the obturator neurovascular plexus; 1, obturator artery; 2, obturator nerve; 3, obturator vein. (**d**) The figure shows the perineum in a female fetus with 35 weeks post-conception; 1, bulbospongiosus muscle; 2, labia minora; 3, urethra; 4, rectum

The urogenital diaphragm and the space between the lower branches of the pubis and the branches of the ischium (Fig. 5.3) are formed by the sphincter of the urethra and the deep transverse muscles of the perineum, which are surrounded by an upper and lower fascia layer of the urogenital diaphragm. The closed space between the superficial and deep layers of the fascia is known as the deep perineal space.

The bulbospongiosus muscles guard the entrance to the vagina just below the pelvic floor and surround the vestibular bulbs anteriorly and Bartholin's glands posteriorly. Medial to Bartholin's gland, the urethrovaginalis sphincter surrounds the vaginal introitus. Well-developed fibrous tissue connects the vaginal wall to the urethra anteriorly, to the deep perineal muscle laterally and its continuation into the perineal body posteriorly (Wu et al. 2020) (Fig. 5.2). The external urethral sphincter is a complex structure with proximal (superior) and distal (inferior) parts. The urethral sphincter complex does not cover the portion of the urethra between vestibular bulbs. The most distal part of the sphincter complex, the thin urethrovaginal sphincter, encircles both urethra and vagina, whereas the more proximal portions posteriorly open loops with the vaginal wall (Zijta et al. 2012; House et al. 2017).

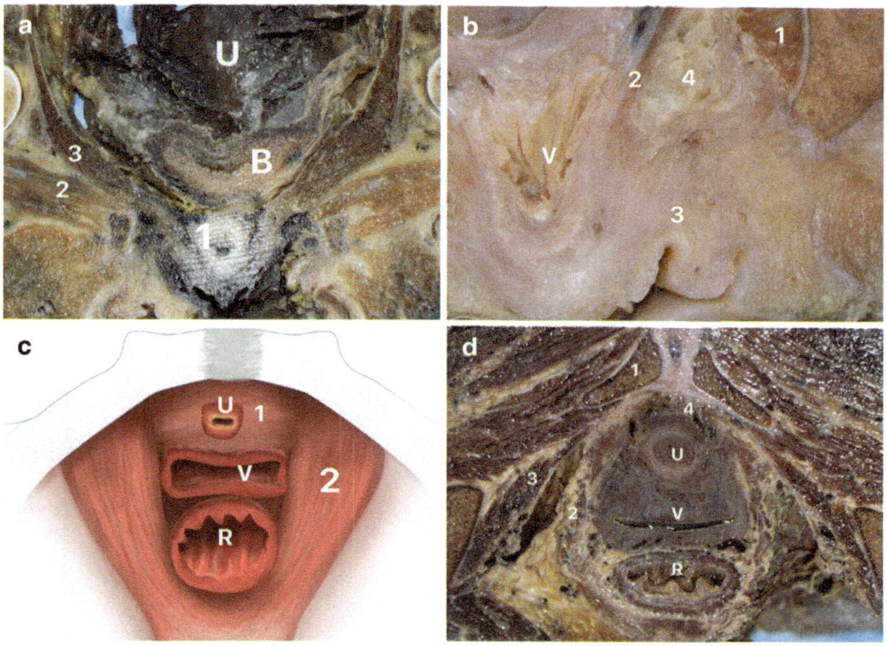

Fig. 5.3 (**a**) The figure shows a frontal section of a fixed corpus showing some details of the female pelvic floor and the relationships between the bladder (B) and the uterus (U); 1, urethra; 2, external obturator muscle; 3, internal obturator muscle. (**b**) The figure shows a frontal section of a fixed corpus showing some details of the superficial perineal space (3); V, vagina; 1, internal obturator muscle; 2, levator anus muscle (LAM); 4, ischiorectal fossa. (**c**) Schematic drawing of a superior view of the female pelvic floor showing some details of pelvic diaphragm and LAM. R, rectum; U, uterus; V, vagina; 1, urethral sphincter; 2, pubococcygeus muscle. (**d**) The figure shows a transverse section of a fixed corpus showing some details of the relationships of the female pelvic viscera; U, urethra; V, vagina; R, rectum; 1, pubis; 2, levator anus muscle (LAM); 3, internal obturator muscle; 4, retropubic space

5.2.3 Pelvic Fascia in Females

The pelvic fascia can be divided into two leaflets: parietal and visceral. The parietal fascia of the pelvis: it is part of a general layer that lines the inner face of the abdominal and pelvic walls. Its continuity with the transversalis fascia and iliac fascia is frequently interrupted by the fusion of these with the periosteum that covers the end lines of the bones and body of the pubic bone (Williams et al. 1995). This fascia can be named according to the muscle it covers (Fig. 5.4). In the lateral walls of the pelvis, it becomes thicker in order to cover the inner face of the anus and coccygeus elevators, thus constituting the upper fascia of the pelvic diaphragm (Rouviere 1968) (Fig. 5.4).

The visceral fascia of the pelvis, also called subperitoneal connective tissue of the pelvis, is formed by the extraperitoneal mesh and serves as an envelope for the

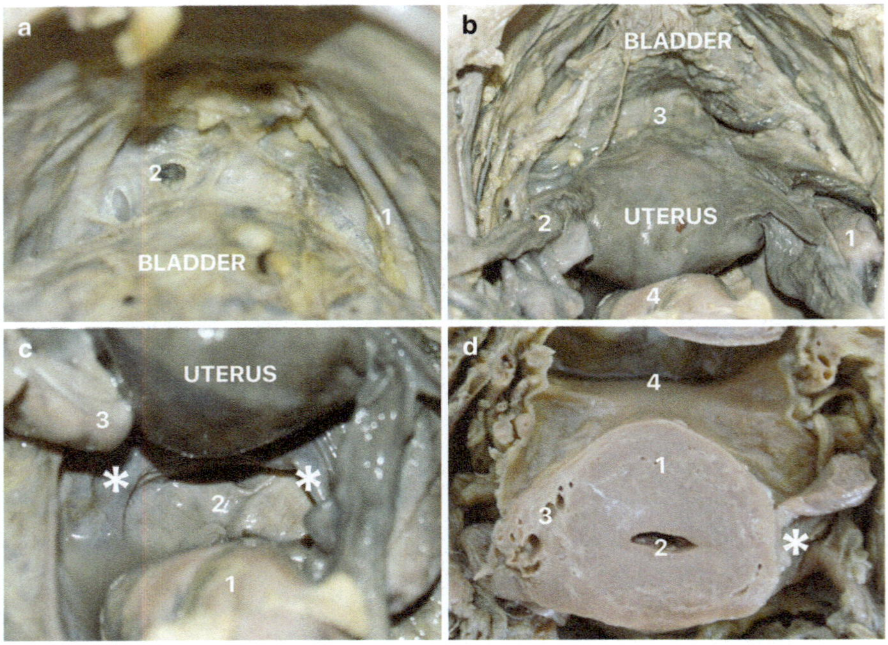

Fig. 5.4 The figure shows the female pelvic fascia and ligaments. (**a**) This figure shows the endopelvic fascia (2) in a female corpus and the pubovesical ligament (1). (**b**) In the same corpus, we can observe the relationships between the uterus and the bladder; 1, right ovary; 2, left uterine tube; 3, vesicouterine space; 4, rectum. (**c**) The figure shows the uterosacral ligaments (*); 1, rectum; 2, rectouterine space; 3, left ovary. (**d**) The figure shows a transverse section at the level of uterine cervix showing the relations of the cardinal ligament (*); 1, uterus; 2, uterine canal; 3, uterine vessels; 4, vesicouterine space

organs and a sheath for the vessels. It is located between the peritoneum and the parietal fascia. It encompasses the pelvic organs and continues with the parietal strip at the points where these organs pass through the pelvic floor. It is in this connective tissue that the uterus, vagina, bladder, and rectum are housed (Williams et al. 1995). This tissue condenses to form the sheaths of the rectum, vagina, and bladder and to form some ligaments. The main ones are the following:

1. Transverse cervical ligaments (or cardinal or Mackenhodt ligaments): they are fibromuscular condensations of the pelvic fascia that pass to the cervix uterine and upper extremity of the vagina, from the lateral walls of the pelvis (Fig. 5.4).
2. Pubocervical ligaments: they are two dense bands of connective tissue that pass to the cervix from the posterior surface of the pubis. They are positioned on each side of the bladder neck, to which they also provide support (pubic-bladder ligaments) (Netter 1978) (Fig. 5.4).
3. Sacrouterine ligaments: they consist of two fibromuscular bands of the pelvic fascia that pass to the cervix and upper end of the vagina from the lower end of

the sacrum, forming two ridges, one on each side of the rectum uterine excavation (Douglas bag bottom) (Fig. 5.4).

Between the ligaments that form the containment device and the pelvic viscera sheath, there is a loose connective tissue, infiltrated with adipose tissue, which performs the function of elastic filling, and is therefore called the elastic containment device of the pelvic viscera (Williams et al. 1995). There is also the fatty body of the ischiorectal fossa, located below the pelvic diaphragm (Figs. 5.1 and 5.3). This fat formation is compressed by the perineal muscles and tightens even more when sitting. Thus, these formations together behave like elastic cushions that contribute to keeping the pelvic viscera in position. The hormonal deprivation that accompanies menopause results in atrophy of the connective and muscular elements of these devices, critical for the support of the pelvic organs (Rud et al. 1980; Buzelin et al. 1992).

5.3 Female Urethra

The urethra consists of a rich vascular sponge surrounded by a layer of smooth muscle and fibroelastic tissue. Functionally, the surrounding smooth muscle layer maintains the continuous direction of the submucosal expansive forces inwardly toward the mucosa (Fig. 5.3b). This efficient mucosal seal is the major contribution to the urethral closure mechanism. This mechanism is under hormonal control, and the lack of estrogen leads to atrophy and replacement of the vascular supply with fibrous tissue. Other causes of the incompetence of the urethral mucous seal are multiple surgeries, trauma, radiation, and neuropathies. When the urethral closure mechanism is lost, stress incontinence appears (Williams et al. 1995).

The local anatomical structures that are important for the continence mechanism can be divided into three parts: distal, proximal, and intrinsic. Additionally, an extrinsic mechanism occurs during sudden increases in abdominal pressure (due to the transfer of this pressure to the intra-abdominal urethra). The urethra is composed of the mucosa, submucosa, periurethral elastic tissue, and urethral muscle (Fig. 5.3). As already seen, the mucosa and submucosa are subject to hormonal stimulation, necessary for the maintenance of the urethral mucous seal. The healthy urethral mucosa is thick and elastic, with a rough inner surface. These factors are very important to obtain an airtight closure of the urethral lumen. The abundant venous channels in the submucosa form a turgid cushion, located between the mucosa and the muscular wall. It has been shown in humans and dogs that the venous plexus contributes approximately 30% to the passive closure mechanism of the female urethra (Tullock 1974; Levin et al. 1980; Klutke 1989).

These mechanisms help ensure an efficient and airtight closure of the urethra with minimal pressure from the muscle wall. The elastic fibers in the urethral wall are abundant and probably have a passive role in affixing the walls of the urethra. When elastic fibers are replaced by fibrous tissue as a result of trauma or irradiation, the urethra becomes rigid and deficient, both in opening and closing. The fourth component is the urethral muscle, that is, the intrinsic muscle of the urethral wall, which includes both smooth and striated muscle. Through its tone, the urethral muscle maintains centripetal pressure over the two layers previously described, closing the urethral lumen. There is experimental evidence that, in addition to the urethral mucosa and submucosal venous plexus, smooth muscle adrenergic receptors are also estrogen-dependent (Levin et al. 1980; DeLancey 1988). Thus, an increase in the level of estrogen will increase muscle tone and pressure of urethral closure.

The bladder neck and the proximal urethra are supported by a set of muscle-fascial structures originating from the levator ani muscles. These structures are called urethro-pelvic ligaments, responsible for maintaining the continence mechanism at the height of the bladder neck (Klutke 1989) (Fig. 5.5). In healthy women, the bladder neck and proximal portion of the urethra are elevated in relation to the lateral wall of the pelvis by these ligaments, which have sufficient strength and elasticity to raise and lower according to changes in intra-abdominal pressure.

5.4 Uterus

The uterus is located in the pelvis, anterior to the rectum and posterior to the bladder (Fig. 5.6), in an anteflexion position in relation to the vaginal axis and flexed in relation to its own axis (Testut 1921; Rouviére 1961). The pelvic viscera in women are covered with peritoneal reflections, which give rise to pouches that have surgical importance. There are three pouches in the female pelvis: (1) prevesical (space of Retzius)—between the pubis and bladder; (2) vesicouterine—between the bladder and uterus; and (3) rectouterine (Douglas pouch)—between the rectum and uterus (Fig. 5.6). The pouch of Douglas is particularly important for being the point of greatest declivity of the female peritoneal cavity when the individual is in orthostatic position, leading to the accumulation of secretions in this space. The pouch of Douglas can be punctured through the vaginal region, which surrounds the lower part of the uterus—the vaginal fornix. Its examination can help in the diagnosis of certain pathologies, such as ectopic pregnancy (Williams et al. 1995) (Fig. 5.7).

The peritoneum that covers the uterus is called the broad ligament, and it is an important structure in the lining and attachment of the organ. Beneath the broad ligament, there is a tissue that has various thick spots that serve to stabilize the pelvic viscera (Williams et al. 1995). This tissue is called the subperitoneal pelvic connective tissue and gives rise to various ligaments. The two most important for attachment of the uterus are the uterosacral ligament and lateral cervical ligament (Fig. 5.4).

The lateral cervical (cardinal) ligament envelops the region just above the uterus cervix and has great surgical importance for being the place where the uterine artery passes before going to the uterus. This is the point where the uterine artery crosses the ureter, an anatomical relationship of great interest when performing hysterectomies (Fig. 5.4) (Netter 1978).

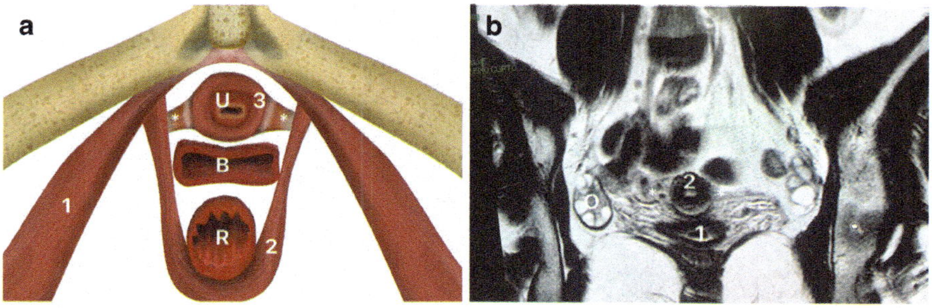

Fig. 5.5 (**a**) Schematic drawing showing the urethro-pelvic ligaments (*) in a transverse section of a female pelvis; U, urethra; B, bladder; R, rectum; 1, internal obturator muscle; 2, levator anus muscle; 3, urethral sphincter. (**b**) Magnetic resonance image (MRI) showing a transverse section of a female pelvis; we can observe the bladder (1), the uterus cervix (2), the ovary (O), and the transverse cervical ligaments (Mackenhodt ligaments) (*)

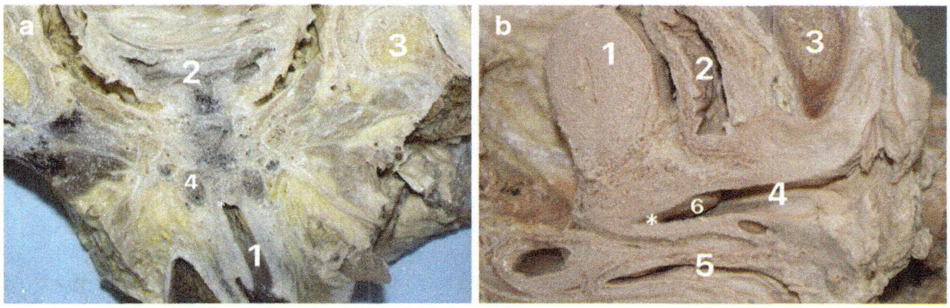

Fig. 5.6 The vaginal fornix. (**a**) The figure shows a frontal section of a female pelvis showing the vaginal relationships; 1, vagina; 2, bladder; 3, femur head; 4, bulbospongiosus muscle; *, vaginal fornix. (**b**) The figure shows a sagittal section of a female pelvis showing the relationships of the uterus and vagina; 1, uterus; 2, bladder; 3, pubis; 4, vagina; 5, rectum; 6, uterine cervix; *, vaginal fornix

Another ligament that is important for uterus attachment, also sectioned during hysterectomy, is the round ligament, a tapered ligament that joins the body of the uterus near the uterine tube to the large lips. This ligament is the main component of the female inguinal canal, and it is the homolog of the gubernaculum in men. The uterus has four parts: fundus, body, isthmus, and cervix. The fundus is the region located above the uterine tubes; the body composes the major portion of the uterus; the isthmus is a narrow region located between the body and cervix; and the cervix is the lowest region of the uterus, enveloped by the vagina (Testut 1921; Rouviére 1961).

The uterine cervix contains the uterine ostium, which establishes the communication between the uterine cavity and vagina (Fig. 5.7). The uterus cervix is the region of greatest clinical concern because it is commonly afflicted by pathologies, such as cervical cancer (Williams et al. 1995).

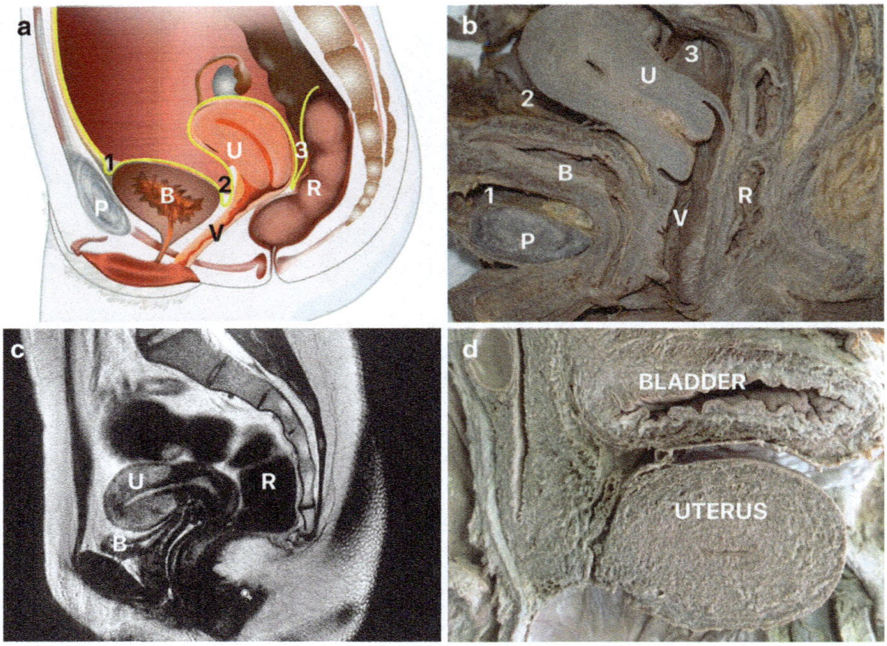

Fig. 5.7 The figure shows a female pelvis with the anatomical relations of the uterus (U) with the rectum (R) and bladder (B). (**a**) Schematic drawing of the sagittal section showing the relations of the uterus with the pelvic viscera; note the peritoneal reflections highlighted in yellow. B, bladder; U, uterus; R, rectum; P, pubis; 1, prevesical space; 2, vesicouterine space; 3, Douglas pouch. (**b**) The figure shows a sagittal section of a formalized female cadaver showing the relations of the uterus with the pelvic viscera. B, bladder; U, uterus; R, rectum; P, pubis; V, vagina; 1, prevesical space; 2, vesicouterine space; 3, Douglas pouch. (**c**) The figure shows a magnetic resonance image of a female pelvis showing the relationships between the uterus (U), bladder (B), and rectum. (**d**) The figure shows a transverse section of a formalized female cadaver showing the relationships between the uterus and bladder

5.5 Uterine Tubes

The uterine (or Fallopian) tubes are bilateral ducts that extend from the uterus to the ovaries, with the function of capturing and conducting the oocyte to the uterine cavity. They permit the spermatozoids to meet the oocyte, so they are the usual place of fertilization and initial division of the ovule (zygote). The tubes are situated in the pelvic cavity, in the upper part and between the two layers of the broad ligament of the uterus. Like the ovaries, they are located approximately 8 to 10 mm below the upper opening (inlet) of the pelvis. The direction of tubes in each hemipelvis is generally lateral, from the uterus to the uterine end of the ovary, so they pass over the mesovaric margin, arching posterolaterally and projecting over the tubal extremity of the uterus and ending on the medial surface and free margin (Kamina 1968).

The tubes are covered by the peritoneum, and its continuation with the pelvic segment occurs through the mesosalpinx, which by homology with other mesos in the abdomen, rests loosely on them, unlike what occurs in the uterus body, where it is adhered to the organ (Fig. 5.7) (Williams et al. 1995). Four morphologically distinct parts are recognized in each tube: the infundibulum, ampule, isthmus, and uterine (intramural) parts.

The uterine tubes are contained in the broad ligament, a peritoneal formation that extends from the pelvis wall to the edge of the uterus. They are surrounded by the peritoneum, whose two layers (anterior and posterior) compose the mesosalpinx. The uterine tubes are mobile in relation to the uterus and pelvic walls, following the movements of the uterus during pregnancy and retroversion (Woodruff and Pauerstein 1969).

Together with the mesosalpinx, the uterine tubes form the upper edge of the broad ligament, which is located posterior in relation to the surface of the round ligament of the uterus and anterosuperior in relation to the ligament of the ovary itself. Between the uterine tubes and the round ligament, which diverge (the former laterally and the latter laterally), is the ovarian recess (Woodruff and Pauerstein 1969). The mesosalpinx contains the vessels and nerves of the uterine tubes, and sometimes embryonic vestiges: epoophoron, located laterally, and paraovarian medially (Woodruff and Pauerstein 1969). Beside the mesosalpinx, the uterine tubes are located below the rings of the small intestine, anterolateral to the rectal ampoule and posterolateral to the bladder, maintaining this relation with these organs when they are full. The ampoule and infundibulum are connected to the pelvic wall in front of the ovary and below the iliac vessels (Williams et al. 1995).

5.6 Ovaries

The ovaries are the pair of primary sexual organs of women, which produce oocytes (ovules) and sexual hormones such as estrogen and progesterone. For this reason, they are considered important organs of the feminine reproductive apparatus and are homologs of the testes in men (Williams et al. 1995).

The ovaries are located in the pelvic cavity, usually 8–10 mm below the upper pelvic opening, in the so-called ovarian recess. Their major axis has superoinferior, lateromedial, and anteroposterior obliquity (Fig. 5.8). They are found in an anterolateral position in relation to the rectum, posterior to the broad ligament and 15–20 mm in front of the sacroiliac joint (Bergman et al. 1988). This location is the result of migration that, like for the testes, starts in the lumbar region medially to the mesonephric duct, but unlike the testes, they are retained in the pelvis definitively, generally in the ninth month of intrauterine life (Williams et al. 1995).

The lateral face of the ovary has relations with the external iliac veins, internal iliac veins, and ureters, separated from them only by the lateral pelvic peritoneum, with which it can present important alterations as a result of varied pathologies (Williams et al. 1995).

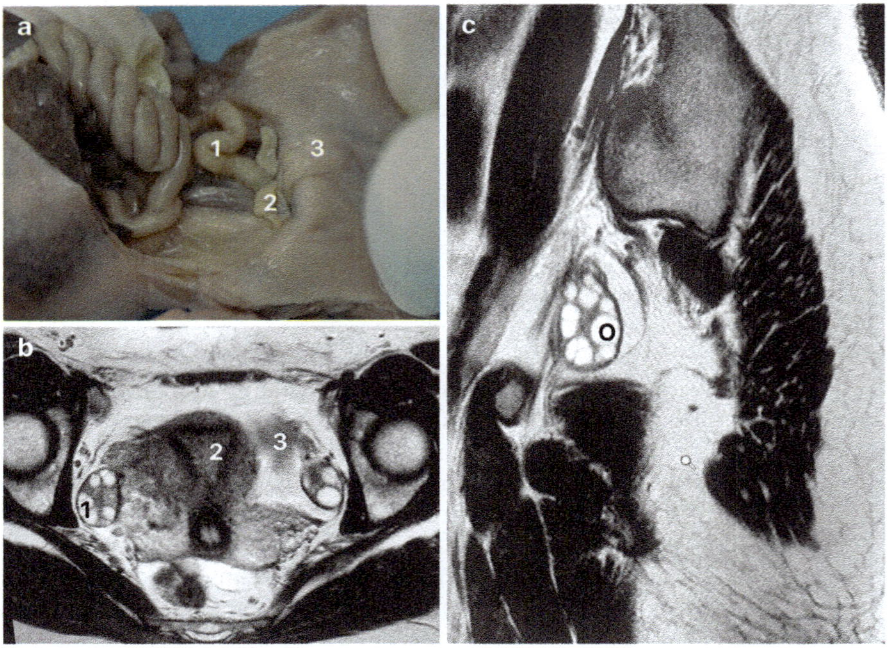

Fig. 5.8 The figure shows some aspects of the ovaries. (**a**) Female fetuses with 22 weeks post-conception; the abdominal wall was opened, and we can observe the ovaries (2); 1, sigmoid colon; 3, bladder. (**b**) The figure shows a transverse section of a magnetic resonance image (MRI) of a female patient; we can observe the left ovary (1), uterus (2), and uterine tube (3). (**c**) The figure shows a sagittal section of an MRI of a female pelvis showing the ovary (O) position

The medial face and free (posterior) edge of the right ovary is connected to the rings of the small intestine, while the left ovary is also connected to the sigmoid. The mesovarium edge, where the blood vessels, lymph nodes, and nerves penetrate and continue accompanying the peritoneum, is one of the attachment points of the ovary, called the ovarian hilum. Each ovary is attached by four ligaments: one on the pelvic wall and three on the ovary itself. These multiple ligaments allow the ovary to move because they converge to the hilum. This enables each ovary to accompany the ascending and descending movements of the uterus during pregnancy and post-partum. Their traction during surgery permits moving them to a position far from the original one (Williams et al. 1995).

5.7 Vascularization, Lymphatic System, and Innervation of Female Pelvis

The main artery responsible for the irrigation of the uterus is the uterine artery, a branch of the anterior trunk of the internal iliac artery. The uterine artery passes over the cardinal ligament and ascends through the lateral regions of the uterus to irrigate it (Fig. 5.9). The uterine artery also participates in irrigation of the uterine tubes, ovary, and part of the vagina. The uterine tube vessels originate from the ovarian and uterine trunks, with the following origins: a) superolaterally from the tubal branch of the ovarian artery (lateral tubal artery) and b) medially from the tubal branch of the uterine artery. The two arteries reach the uterine tube (Koritké et al. 1967; Borell and Fernström 1953). The main arterial irrigation of the ovary is performed by the ovarian artery and right branch of the abdominal aorta artery, arising from the anterolateral contour of the aorta just below the renal arteries, at the level of the L2/L3 intervertebral disk (Machnicki and Grzybiak 1999).

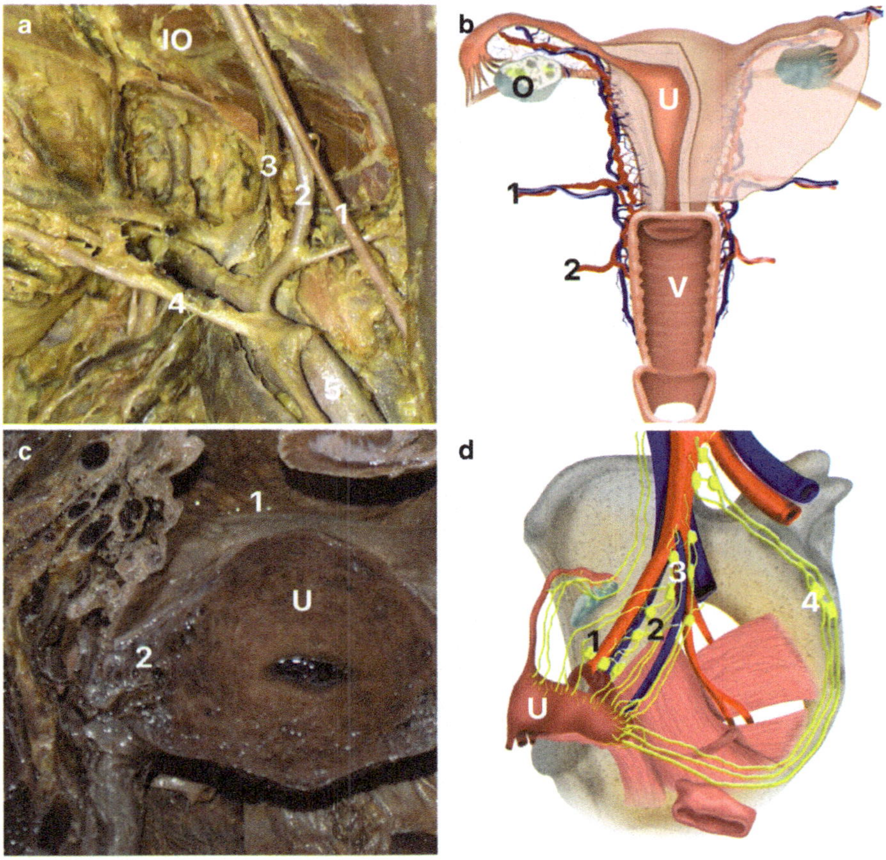

Fig. 5.9 Arterial and lymphatic drainage of a female pelvis. (**a**) Sagittal section of a female pelvis showing some branches of the iliac intern artery (5) and the obturator neurovascular structures; 1, obturator nerve; 2, obturator artery; 3, obturator vein; 4, uterine artery; IO, internal obturator muscle. (**b**) Arterial irrigation of the uterus; note that the uterine artery (1) is also responsible for the irrigation of part of the uterine tube and ovary (O); U, uterus; V, vagina; 2, vaginal artery. (**c**) The figure shows a transverse section of a female pelvis in a fixed corpus; we can observe the vessels (2) that irrigates the uterus (U); 1, rectouterine space. (**d**) Schematic drawing showing the lymphatic drainage of the uterus and vagina to external iliac lymph nodes (1), obturator lymph nodes (2), internal iliac lymph nodes (3), and sacral lymph nodes (4)

The venous drainage is performed by the uterine veins, which accompany the arteries in the direction of the internal iliac vessels (Williams et al. 1995). The uterine lymphatic drainage is of great importance in cancer surgeries. The upper part of the uterine body drains to the lumbar lymph nodes; the lower part of the body to the external iliac lymph nodes; the cervix drains to the internal iliac, external iliac, and sacral lymph nodes; and the region near the round ligament drains to the superficial inguinal lymph nodes (Sampson 1937) (Fig. 5.9).

The nerve supply, mainly along the ovarian and uterine arteries, follows a similar pattern of distribution. The greater part of each tube has sympathetic and parasympathetic innervation. The vagal nerve fibers reach the lateral half, while the pelvic-splanchnic fibers reach the medial half. The sympathetic innervation comes from the tenth thoracic segment and from the second lumbar spinal segment (Chiara 1959; Damiani and Capodacqua 1961). The afferent nerve fibers accompany the sympathetic motor innervation, penetrating the spinal marrow through the corresponding dorsal roots. The postganglionic sympathetic fibers and visceral afferent fibers, as well as the preganglionic parasympathetic fibers, are located in the tubal walls.

References

Bergman RA, Thompson SA, Afifi AK, Saadeh FA. Compendium of human anatomic variation. Baltimore: Urban & Schwarzenberg; 1988.

Borell U, Fernström I. Adnexal branches of uterine artery; arteriographic study in human subjects. Acta Radiol. 1953;40:561–82.

Buzelin JM, Glemain P, Minaire P: The physiology of continence and te physiopathology of urinary incontinence. In: Steg A (ed), Urinary incontinence, Churchill Livingstone, Edinburgh, p13, 1992.

Chiara F. Study of the fine innervation of the female genitália. I. Uterus. Annali Ostet Ginecol. 1959;81:553–76.

Damiani N, Capodacqua A. On the intrinsic innervation of the fallopian tube. Annali Ostet Ginecol. 1961;83:436–46.

DeLancey JO. Structural aspects of the extrinsic continence mechanism. Obst Gynecolo. 1988;72:296.

House M, Bhadelia RA, Myers K, Socrate S. Magnetic resonance imaging of three-dimensional cervical anatomy in the second and third trimester. Eur J Obstet Gynecol Reprod Biol. 2009;144(Suppl 1):S65–9. https://doi.org/10.1016/j.ejogrb.2009.02.027.

Kamina P. Anatomie gynecologique et obstetricale. Paris: Librairie Maloine; 1968.

Klutke CG. The anatomy of stress incontinence. American Urological Association Today, v. Sept–Oct, p. 22, 1989.

Koritké JG, Gillet JY, Pietri J. Les artères de la trompe uterine chez la femme. Arch Anat Histol Embryol. 1967;50:47–70.

Levin RM, Shofer FS, Wein AJ. Estrogen induced alteration in the autonomic responses of the rabbit urinary bladder. J Pharmacol Exp Ther. 1980;215

Machnicki A, Grzybiak M. Variations in ovarian arteries in fetuses and adults. Folia Morphol (Warsz). 1999;58:115–25.

Netter F. Reproductive system, vol. 2. New Jersey: Ciba; 1978.

Rouviére H. Anatomía humana descriptiva y topográfica. 8th ed. Madrid: Bailly-Bailliere; 1961.
Rouviére H. Anatomía humana descriptiva y topográfica. 8th ed. Madrid, Bailly-Bailliere, 1968.
Rud T, Andersson KE, Asmmussen M, Hunting A, Ulmsten U. Factors maintaining the intra urethral pressure in women. Investig Urol. 1980;17:343.
Sampson JA. Lymphatics of mucosa of fimbriae of the fallopian tube. Am J Obstet Gynecol. 1937;33:911–30.
Testut L. Anatomíe humaine: appareil uro-génital. Livre X. 7th ed. Paris: Librairie Octave Doin; 1921.
Tullock AGS. The vascular contribution to intraurethral pressure. Br J Urol. 1974;46:659.
Williams PL, Bannister LH, Berry MM, Dyson M, Dussek JE, Ferguson MWJ. Gray's anatomy. 38th ed. Churchill-Livingstone: Edinburgh; 1995.
Woodruff JD, Pauerstein CJ. The fallopian tube. Baltimore: Williams & Wilkins; 1969.
Wu Y, Hikspoors JPJM, Mommen G, et al. Interactive three-dimensional teaching models of the female and male pelvic floor. Clin Anat. 2020;33:275–85.
Zijta FM, Lakeman MME, Froeling M, et al. Evaluation of the female pelvic floor in pelvic organ prolapse using 3.0-Tesla diffusion tensor imaging and fibre tractography. Eur Radiol. 2012;22:2806–13.

Index

L. A. Favorito, N. T. Logsdon (eds.), *Atlas of Sectional Anatomy*,
https://doi.org/10.1007/978-3-030-91688-6

The manufacturer's authorised representative in the EU is Springer Nature Customer Service Centre GmbH, Europaplatz 3, 69115 Heidelberg, Germany. If you have any concerns regarding our products, please contact ProductSafety@springernature.com

Printed and bound by CPI Group (UK) Ltd, Croydon, CR0 4YY
28/09/2026
02224077-0014